STEAM AROUND DONCASTER IN THE 1960s

Keith W. Platt

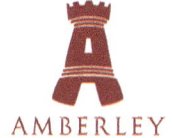
AMBERLEY

First published 2022

Amberley Publishing
The Hill, Stroud
Gloucestershire, GL5 4EP

www.amberley-books.com

Copyright © Keith W. Platt, 2022

The right of Keith W. Platt to be identified as the Author of this work has been asserted in accordance with the Copyrights, Designs and Patents Act 1988.

ISBN 978 1 3981 0546 1 (print)
ISBN 978 1 3981 0547 8 (ebook)

All rights reserved. No part of this book may be reprinted or reproduced or utilised in any form or by any electronic, mechanical or other means, now known or hereafter invented, including photocopying and recording, or in any information storage or retrieval system, without the permission in writing from the Publishers.

British Library Cataloguing in Publication Data.
A catalogue record for this book is available from the British Library.

Origination by Amberley Publishing.
Printed in the UK.

Introduction

I was born and grew up in Thorne, a town situated 10 miles to the north-east of Doncaster. My childhood, in the 1950s, was very much the same as many other post-war children. We had lots of freedom to enjoy outdoor activities of all kinds and were encouraged to participate in different hobbies. From my bedroom window I could glimpse the trains as they made their way along the Doncaster to Hull railway line and my interest in steam locomotives was firmly established. Along with several friends I became a trainspotter, collecting the numbers of all the locomotives and carefully underlining them in my second-hand Ian Allan Eastern Region book. Very soon the lure of the East Coast main line, with the sleek A4s and other Pacific classes, could not be ignored. The 7-mile bike ride to the tiny hamlet of Moss, the closest location of the main line, became a regular feature of our summer days.

A seventeen-minute train journey to Doncaster station opened up a whole new world for us to be part of and we embraced the adventure wholeheartedly.

Doncaster was, and still is, a very important railway junction with the Kings Cross to Edinburgh main line being intersected by connecting lines to Leeds, Sheffield, Hull, Grimsby, and Lincoln. The movement of trains and locomotives was fairly constant throughout the day with certain highlights at different times. The 'shed stream' and the 'Plant stream' were not to be missed as batches of locomotives were taken to the works for overhaul, and shiny locos returned after being repaired and repainted. The cries of 'Streak, streak!' from the expectant crowd of trainspotters lining the platforms preceded the thrill of seeing non-stop expresses rushing through the station.

Doncaster was at the centre of a large coal mining area, which had over fifty collieries within a few miles of the town. Coal trains were a significant part of the railway scene in the 1960s. Much of the heavy freight traffic used avoiding lines to reduce movements through the station area.

On some visits to Doncaster it was possible to attach ourselves to groups of enthusiasts who were visiting the motive power depot and the Plant. This enabled a whole new area of the world of steam locomotives to be explored and enjoyed.

My book is a personal recollection of the steam scenes that were everyday occurrences in the Doncaster area in the 1960s. It includes images at Thorne, Goole, Mexborough, Retford and Scunthorpe, along with various National Coal Board and Steelworks sites. These were all places I include in the Doncaster area because they were just a bike ride away from my home.

All the images in the book are original slides and negatives from my collection, with the photographers credited when known. I would like to thank my brother John B. Platt for his proofreading work and my wife Andrea for her IT expertise, support and advice.

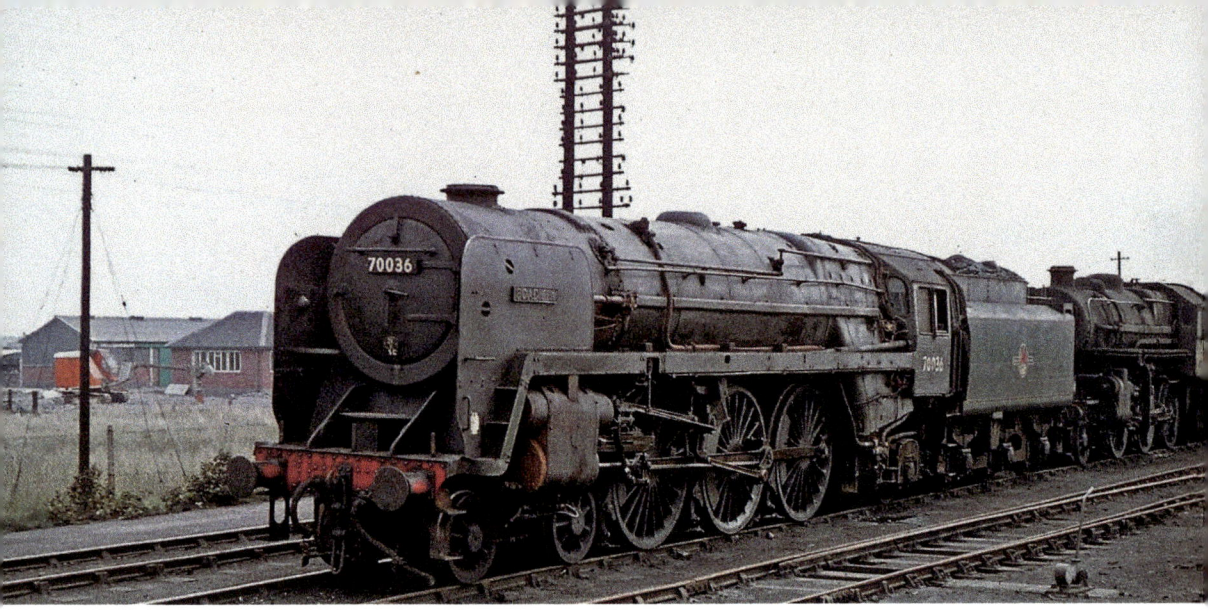

19 February 1961. BR Britannia 4-6-2 No. 70036 on Doncaster MPD.

BR Britannia 4-6-2 No. 70036 *Boadicea* awaits its next duty on Doncaster MPD. It was allocated to March MPD at the time and was a regular visitor to Doncaster on cross-country workings northwards from East Anglia. It was looking in fairly worn condition and was to be reallocated to Immingham MPD before the end of 1961 and also receive a general overhaul at the Plant. Its association with Doncaster ceased in December 1963 when, along with other Immingham-based Britannia locomotives, it was transferred to Carlisle. It was finally withdrawn in October 1966 with a service life of a few months short of fourteen years. As with many other Carlisle-allocated locos it was broken up by Motherwell Machinery and Scrap of Wishaw.

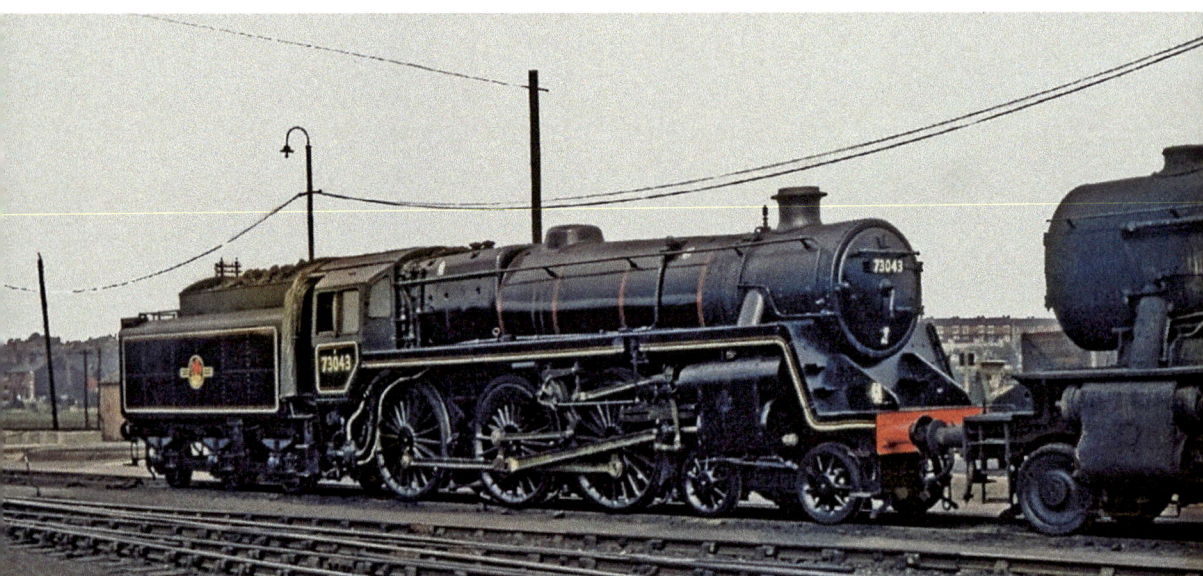

20 August 1961. BR Standard 5MT 4-6-0 No. 73043 on Doncaster MPD.

BR Standard 5MT 4-6-0 73043 had just returned from the Plant Works where it had received a general overhaul and was about to return to its home shed of Canklow. Within eighteen months, however, it would be transferred to the Southern Region and would be allocated to Nine Elms MPD when steam on the Southern finally came to an end in July 1967. The Plant had built forty-two of the class and it received many of the class for general overhauls, including Western Region locomotives that were painted in lined green livery.

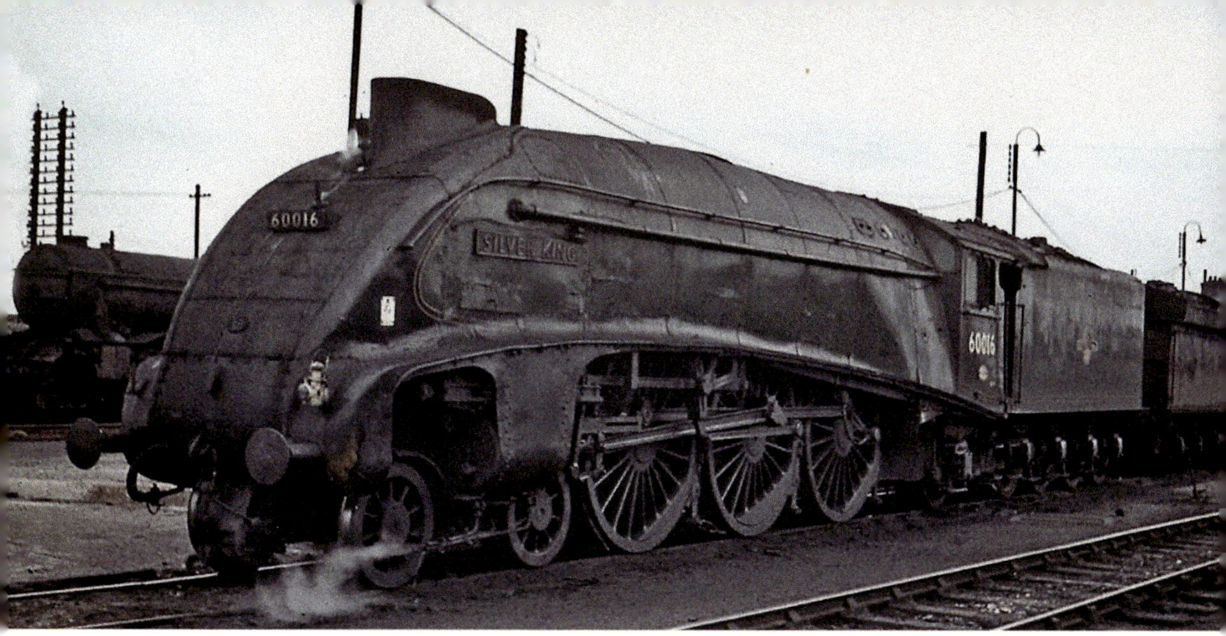

April 1962. LNER 4-6-2 A4 No. 60016 on Doncaster MPD.

LNER 4-6-2 A4 No. 60016 *Silver King* awaits its next duty northbound on Doncaster MPD. It would return to Doncaster the following month to enter the Plant for its last general overhaul. No. 60016 was one of the original batch of four A4s built in 1935 to head the new Silver Jubilee streamlined train. It spent most of its thirty-five-year career based in the north-east of England, but was transferred to Scotland for its last eighteen months of service, working the three-hour expresses between Aberdeen and Glasgow. It was withdrawn in March 1965 from Aberdeen Ferryhill MPD.

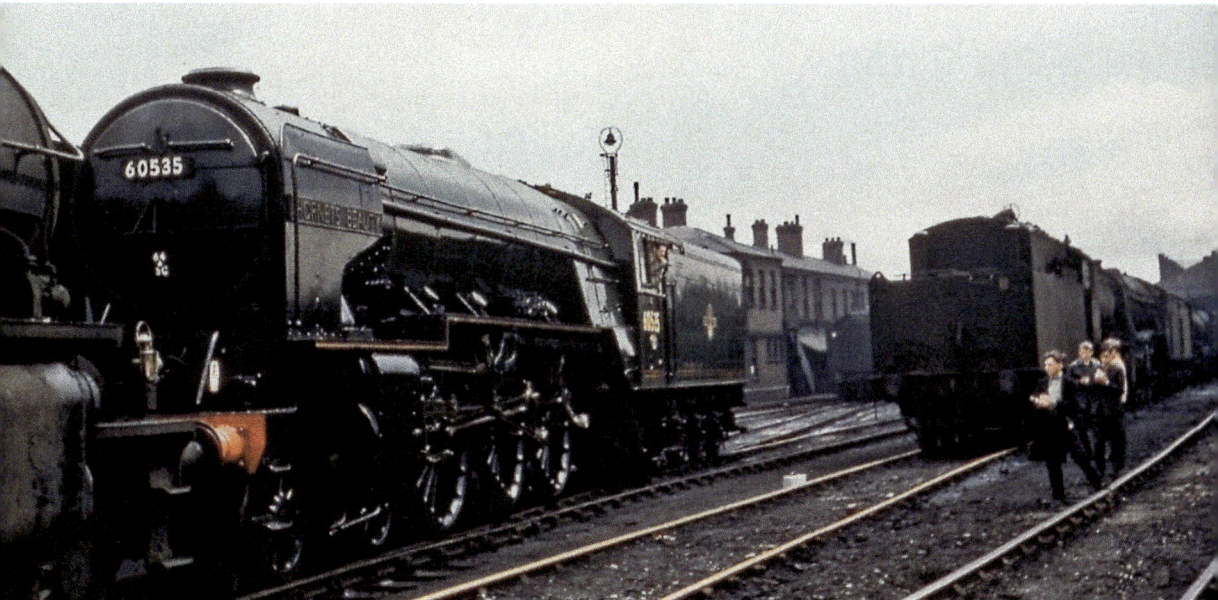

6 May 1962. LNER 4-6-2 A2 No. 60535 on Doncaster MPD.

LNER 4-6-2 A2 No. 60535 *Hornet's Beauty* had been returned to Doncaster MPD from the Plant after its final general overhaul. It was soon to return to Scotland for three more years of service before withdrawal. Gleaming locomotives, fresh from overhaul, were always a source of excitement for the onlooking trainspotters, particularly Scottish Pacifics, which were rarely seen so far south.

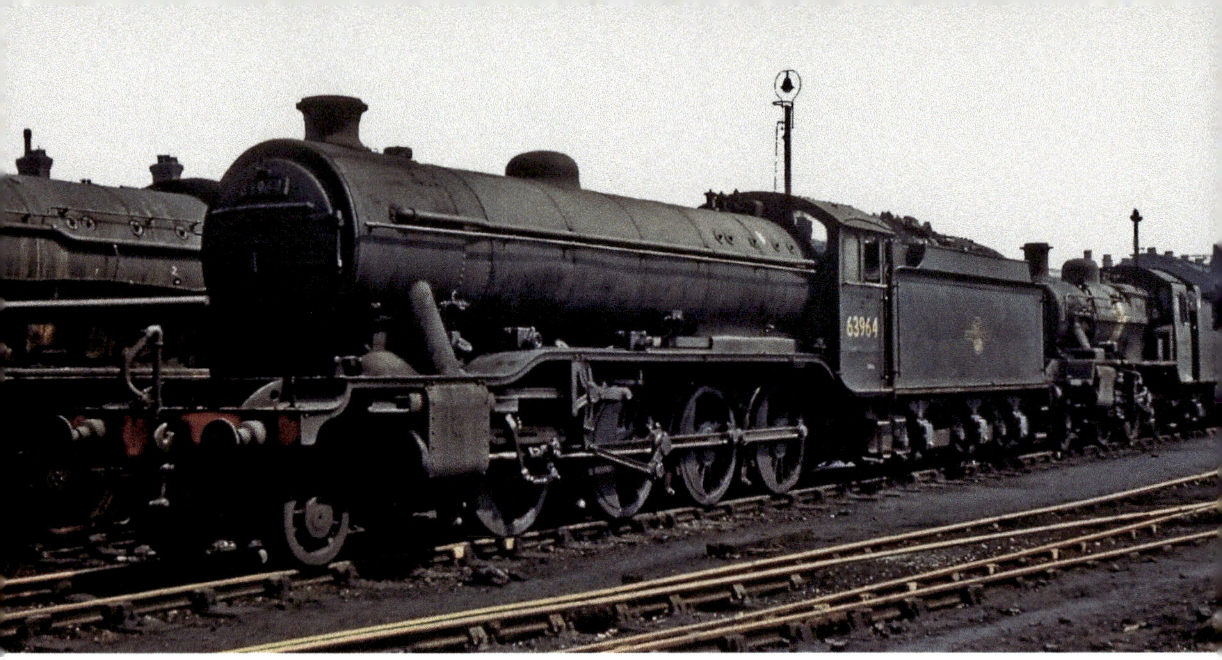

1 July 1962. LNER 2-8-0 O2/3 No. 63964 on Doncaster MPD.

LNER 2-8-0 O2/3 No. 63964 was one of sixty-seven Gresley-designed heavy freight locomotives built in batches between 1921 and 1943. They were used on the heavy iron ore and coal trains, which were so much a part of railway operations in the area. No. 63964 entered service in June 1942 and was eventually withdrawn from Retford MPD in September 1963, a few weeks before the last of the class.

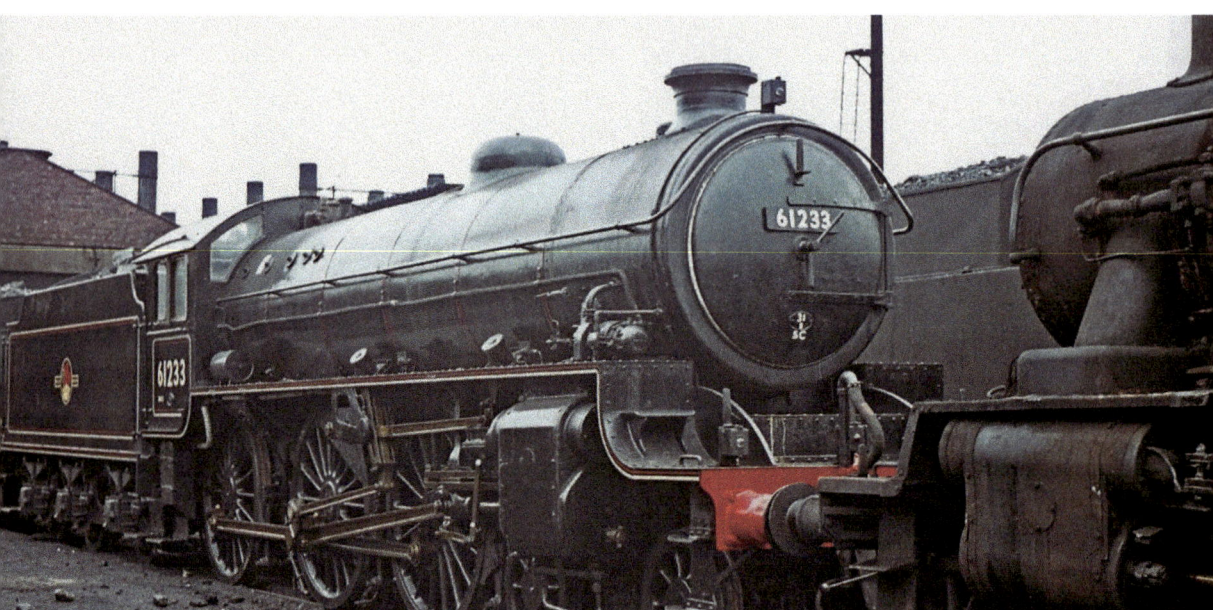

15 July 1962. LNER 4-6-0 B1 No. 61233 on Doncaster MPD.

LNER 4-6-0 B1 No. 61233 was seen on Doncaster shed after a fifty-two-day visit to the Plant for a general overhaul. Amazingly, within two months of its return to its home depot of March it was withdrawn, then reinstated to traffic a couple of weeks later. By December 1963 its services were no longer required and it was transferred to Departmental Stock as No. 21 for use as a stationary boiler for train heating purposes. It was finally withdrawn from these duties in May 1966 from Cambridge and was scrapped.

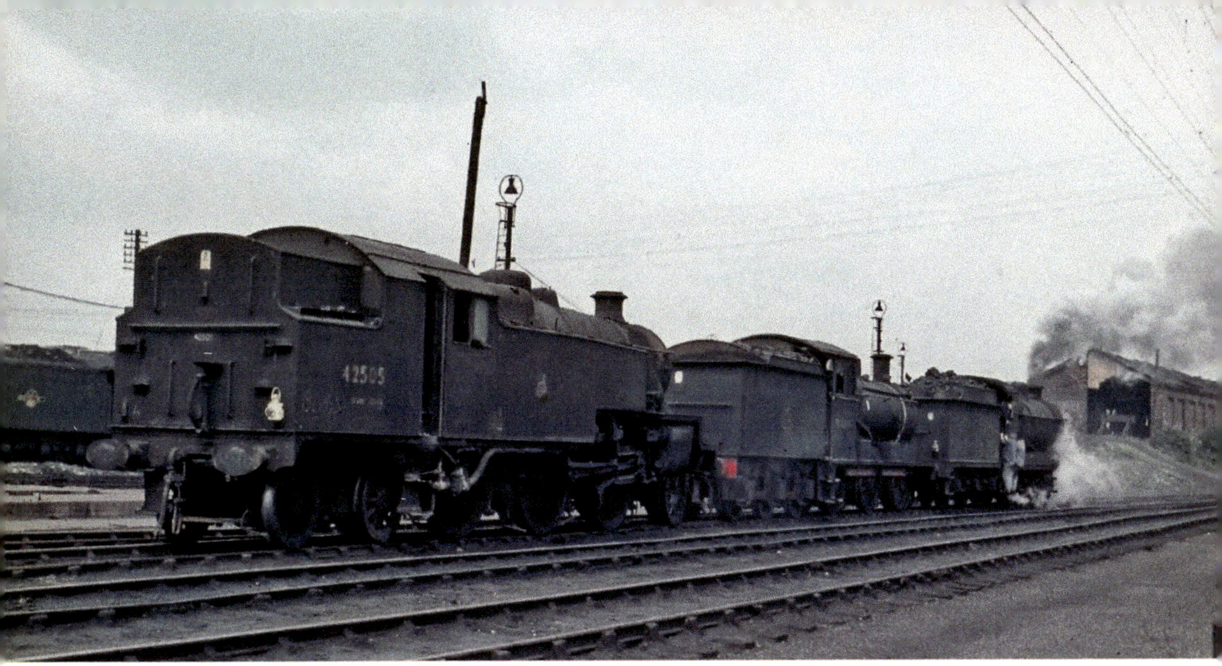

August 1962. LMS 2-6-4T No. 42505 with No. 65567 on Doncaster MPD.

LMS 2-6-4T 4MT No. 42505 and LNER 0-6-0 J17 No. 65567 were coupled together to take the short journey across to the Plant. It would be the last journey to the cutter's torch for the three-cylinder Stanier tank engine, which had been made redundant by the electrification of the Tilbury line. The J17 was to have a much different fate, as it became part of the National Collection of steam locomotives. It was cosmetically restored at the Plant before being displayed at various sites over the intervening years, most recently at Barrow Hill Roundhouse. (L. Flint)

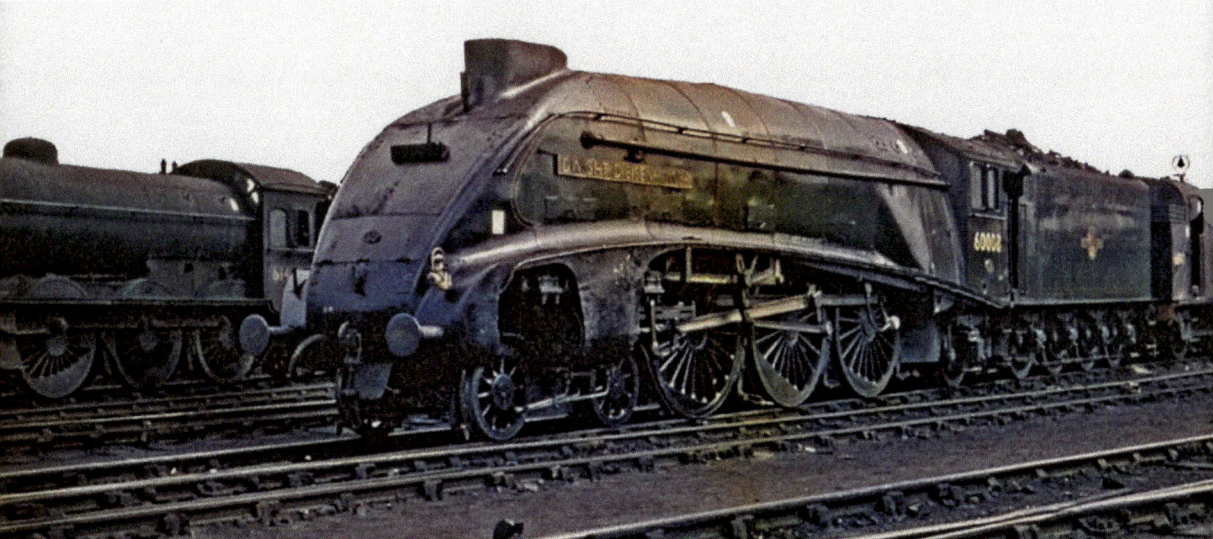

20 August 1961. LNER 4-6-2 A4 No. 60008 on Doncaster MPD.

LNER 4-6-2 A4 No. 60008 *Dwight D. Eisenhower* was a Kings Cross-allocated locomotive that had just been released from the Plant Works after receiving a light casual overhaul. It has been fully coaled and watered ready to return to its work on the East Coast Main Line. Within two years this work had been taken over by diesel locomotives and No. 60008 was returned to Doncaster after its withdrawal from service. It was restored cosmetically and was donated to the United States of America. The locomotive was shipped there on 27 April 1964 and housed in the National Railroad Museum in Green Bay, Wisconsin.

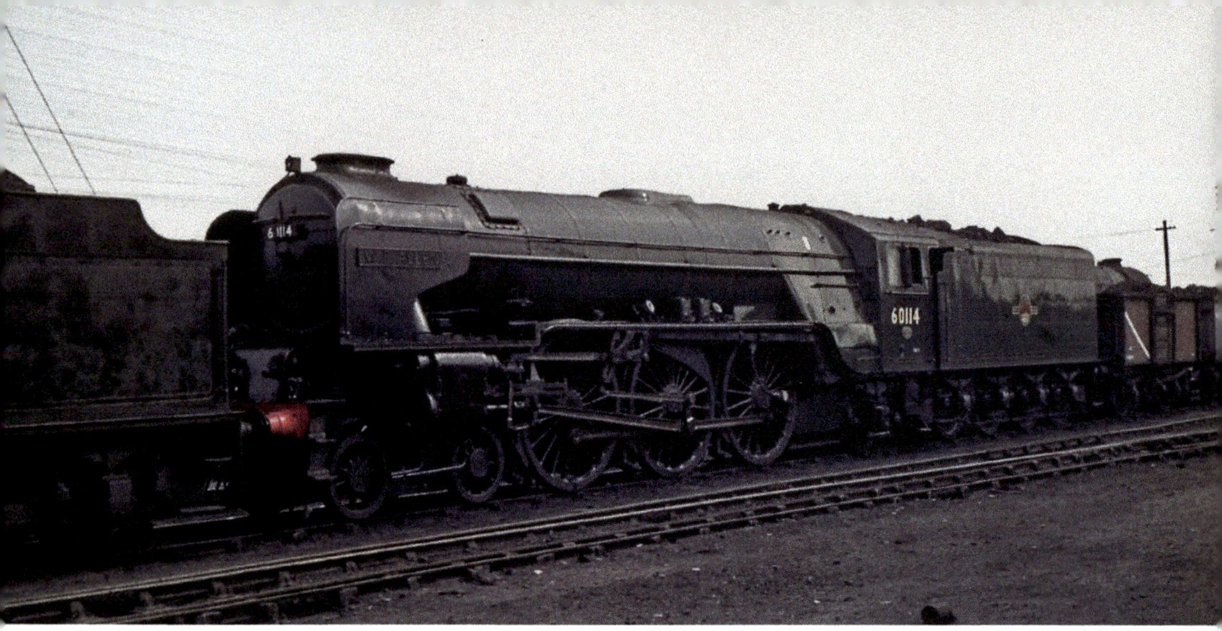

25 November 1962. LNER 4-6-2 A1 No. 60114 on Doncaster MPD.

LNER 4-6-2 A1 No. 60114 *W. P. Allen* still looking fresh from overhaul a few months earlier. No. 60114 was the first of the class, and had been designed under LNER regime by A. H. Peppercorn but constructed by the nationalised British Railways. It emerged from the Plant in July 1948 and became a long-term resident of Doncaster MPD. It was to remain in service for a further two years, being withdrawn in late December 1964 and scrapped by Hughes Bolckow scrapyard in Blyth, Northumberland, in March 1965.

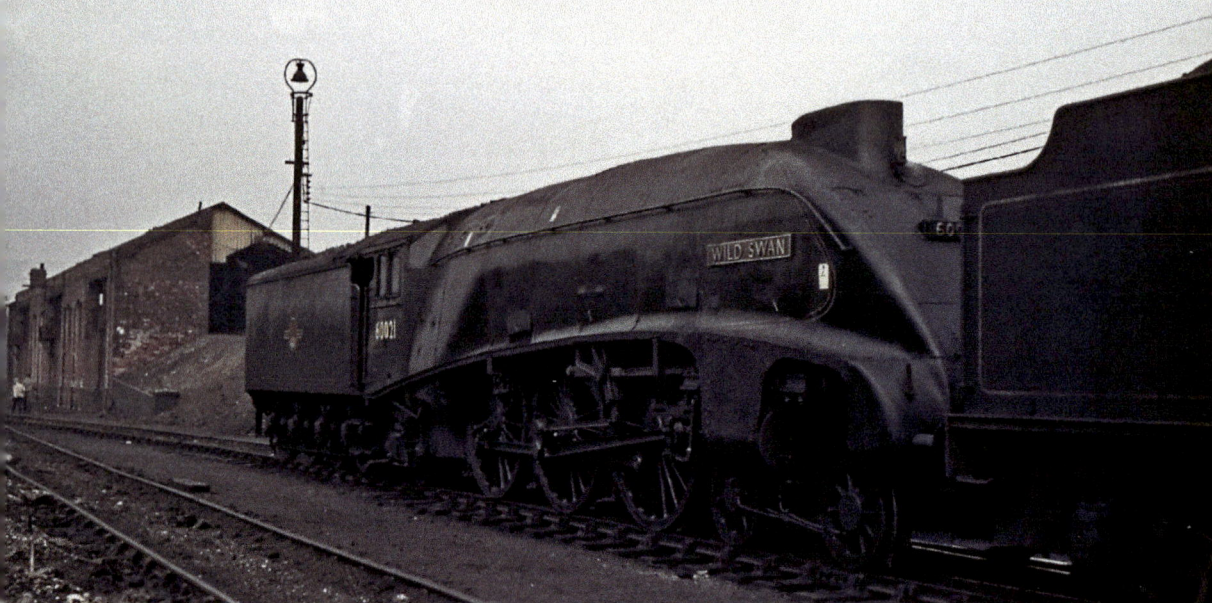

25 November 1962. LNER 4-6-2 A1 No. 60021 on Doncaster MPD.

LNER 4-6-2 A1 No. 60021 *Wild Swan* entered service in February 1938 and spent most of its working life allocated to Kings Cross, with a few short spells at other East Coast Main Line depots. With the closure of Kings Cross shed it spent the last couple of months at New England, being withdrawn from there in October 1963 and cut up at the Plant in early 1964. *Wild Swan* had been a regular and familiar sight in Doncaster over the years, but days before its withdrawal I photographed it on Derby shed after it had worked in with a Newcastle to Bristol train.

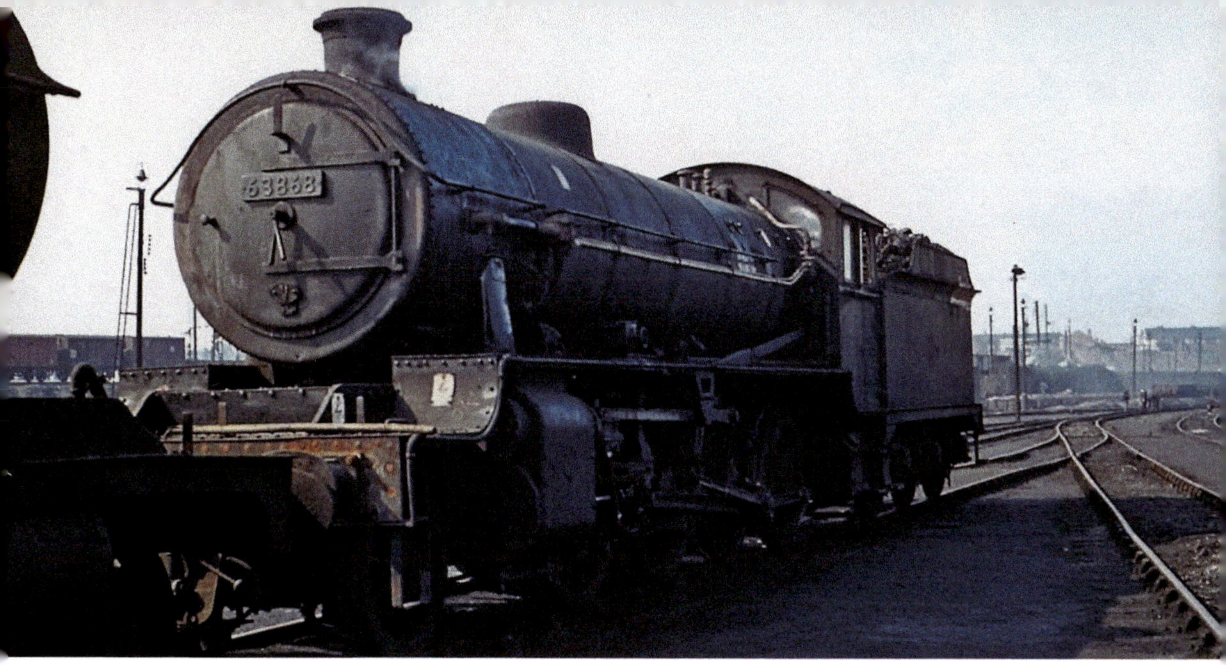

March 1963. LNER 2-8-0 O1 No. 63868 on Doncaster MPD.

LNER 2-8-0 O1 No. 63868 was seen on Doncaster MPD after receiving a heavy casual overhaul at the Plant. This work included the fitting of a replacement boiler but not a repaint. Although it sported a 31B March MPD shed plate, it was to move on to 41H Staveley GC MPD when released to traffic. It was originally built for the Great Central Railway in 1919 as an O4/3, but was rebuilt with new cylinders, boiler and cab in February 1945 as an O1. It was to remain in service until July 1965 and was eventually cut up by Draper's of Hull in November 1965.

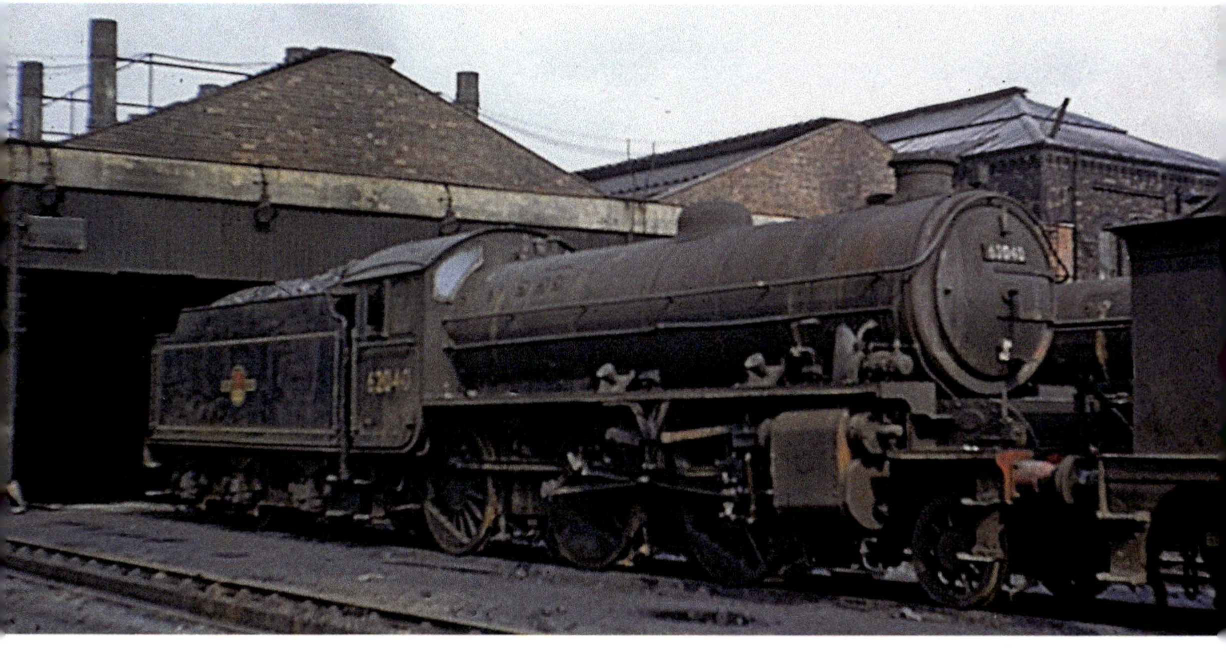

31 March 1963. LNER K1 2-6-0 No. 62040 on Doncaster MPD.

LNER K1 2-6-0 No. 62040 had been reallocated to Doncaster MPD after receiving a general overhaul at the Plant in November 1962. During the few months since then it appears to have received no attention from the cleaners. It was to remain at Doncaster until it was withdrawn in January 1965 and cut up by Draper's of Hull in June of that year.

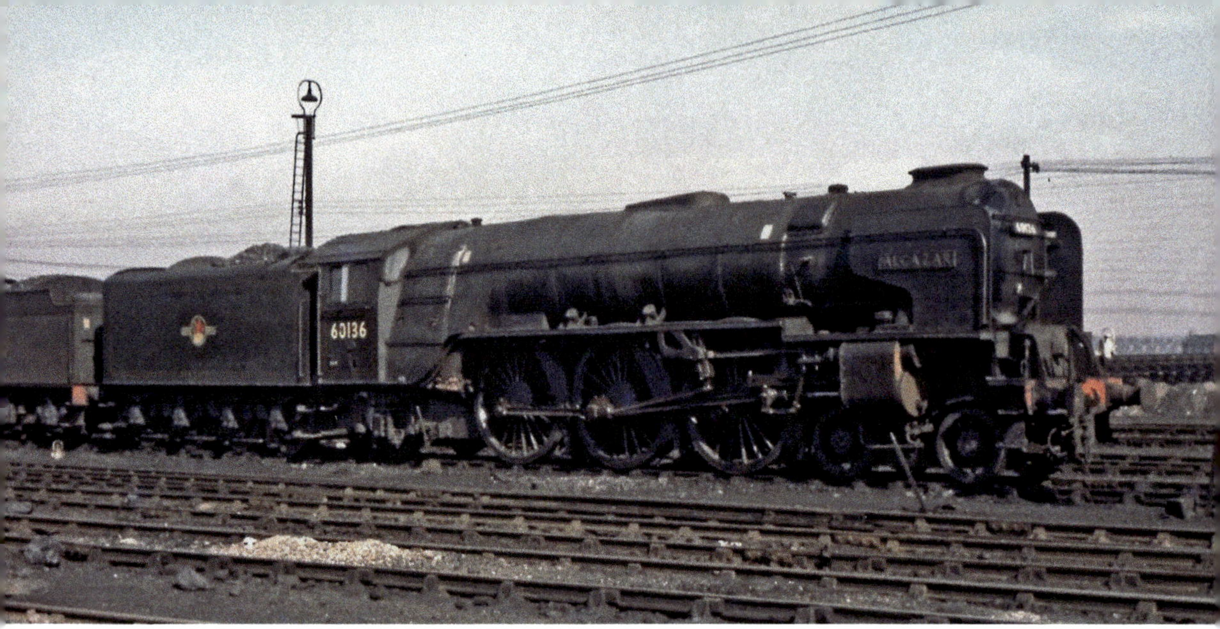

April 1963. LNER 4-6-2 A1 No. 60136 at Doncaster MPD.

LNER 4-6-2 A1 No. 60136 *Alcazar* at rest between duties on its home shed. It has been serviced, coaled and watered and it simmers gently awaiting its next northbound working. Peppercorn A1s were a familiar sight on Doncaster MPD with a good number allocated there over the years and other York, Leeds and Newcastle-based members of the class were frequent visitors. The depot was occasionally the host to Scottish A1s that had been overhauled at the nearby Plant.

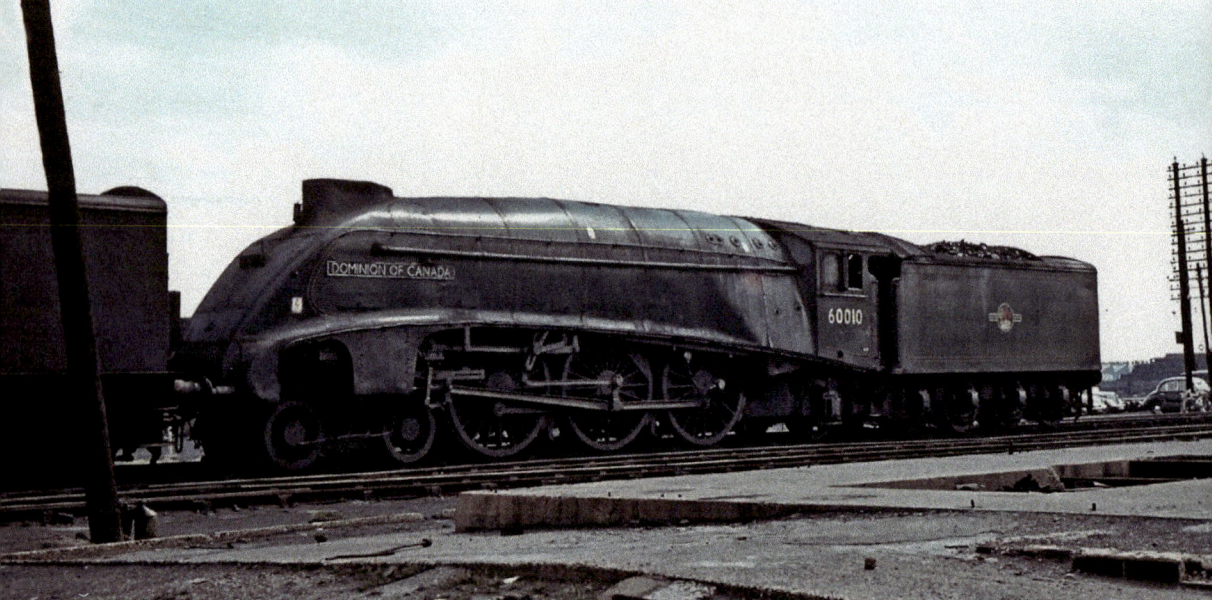

May 1963. LNER A4 4-6-2 No. 60010 on Doncaster MPD.

LNER 4-6-2 A4 No. 60010 *Dominion of Canada* awaits movement to the Plant for a light casual overhaul before its transfer to New England and later in the year to Aberdeen. It worked the Aberdeen to Glasgow services for two years before being sent to Darlington Works for overhaul. The condition of its boiler meant that it was withdrawn and it languished at Darlington for over a year. It was eventually cosmetically restored at Crewe Works in 1966 and shipped to Canada in April 1967 for preservation.

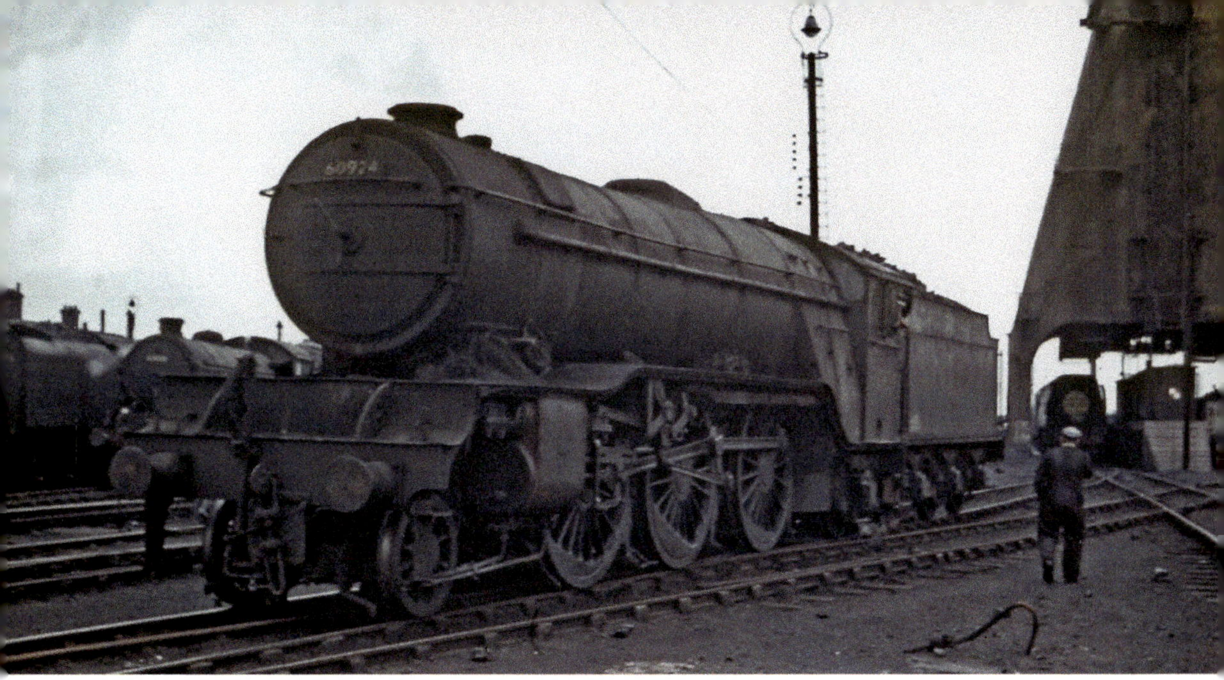

12 May 1963. LNER V2 2-6-2 No. 60924 on Doncaster MPD.

LNER V2 2-6-2 No. 60924 had been allocated to Doncaster MPD for a year, but its stay was to be short-lived as it was withdrawn in September 1963. Just visible in the background under the coaling plant was SR West Country 4-6-2 No. 34094 *Mortehoe*, which had arrived at Doncaster with a Warwickshire Railway Society special from Birmingham New Street.

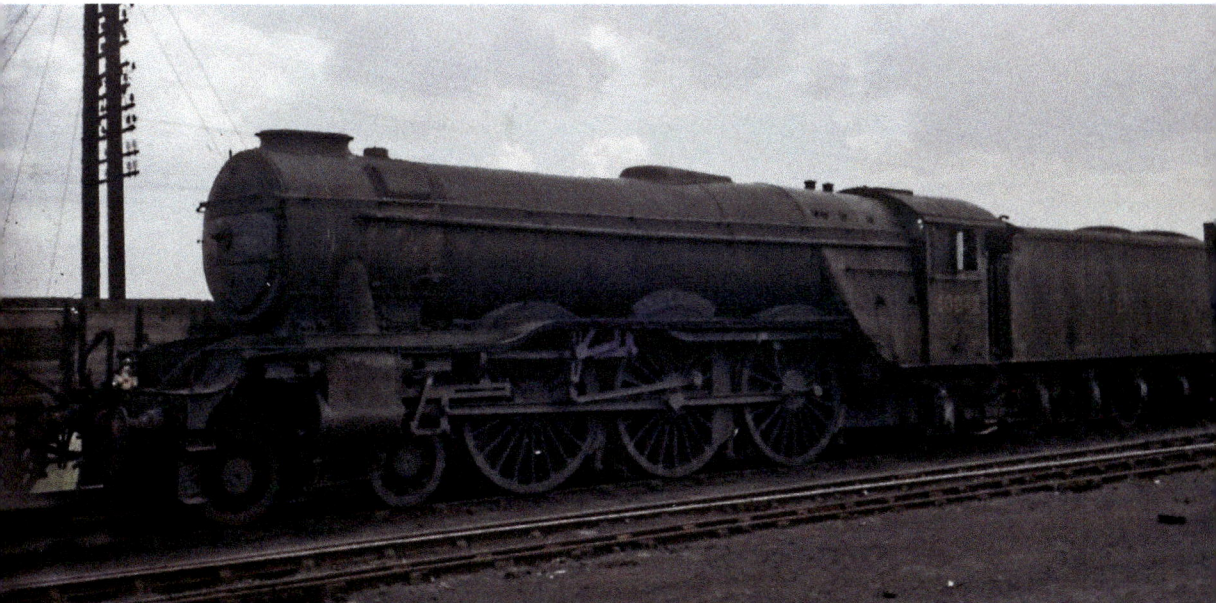

12 May 1963. LNER 4-6-2 A3 No. 60072 on Doncaster MPD.

LNER 4-6-2 A3 Pacific No. 60072 *Sunstar* stands withdrawn on Doncaster MPD. It was awaiting its last short journey across to the Plant for scrapping. It was a locomotive closely associated with the north-east for much of its career until its transfer to Leeds, frequenting the Settle & Carlisle line for almost a year. In July 1962 it returned to Heaton MPD for its final year in service.

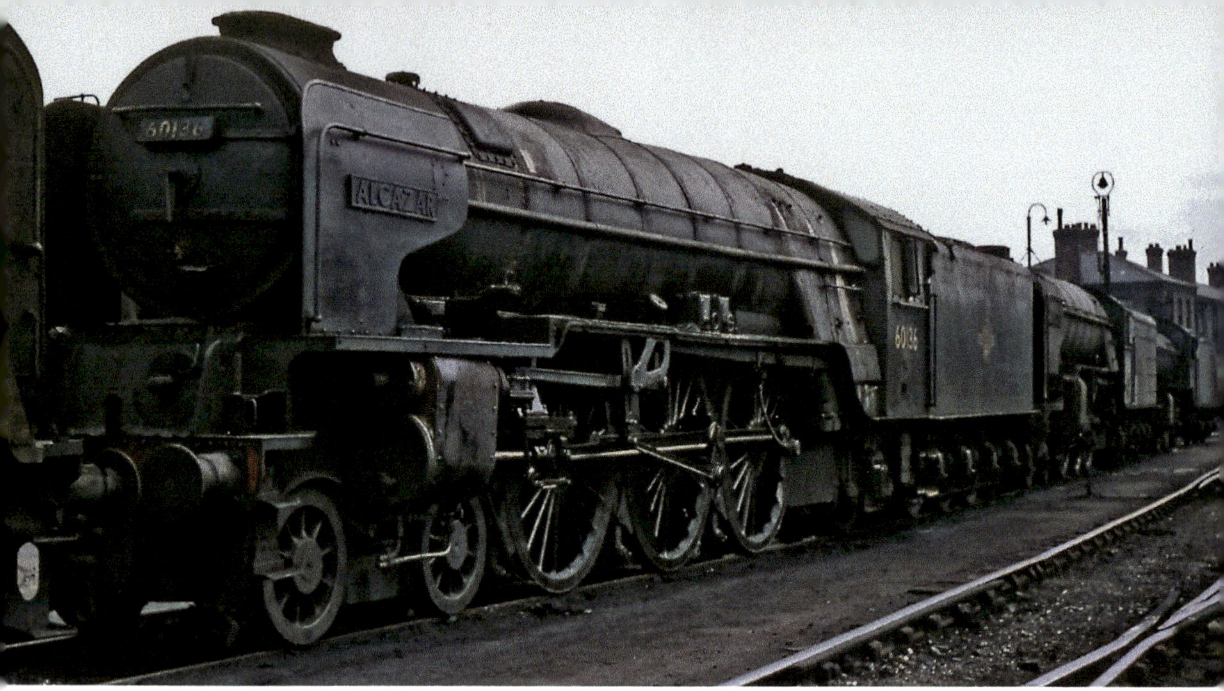

12 May 1963. LNER 4-6-2 A3 No. 60136 at Doncaster MPD.

LNER 4-6-2 A3 No. 60136 *Alcazar* had been a Doncaster-based locomotive since early in 1959, but was to be withdrawn a couple of days after this photograph had been taken. It had been in service a little over fourteen years before dieselisation had caught up with it and it was taken to the Plant for cutting up. During my regular visits to Doncaster over the years *Alcazar* seemed to be ever present and would inevitably turn up at some point in the day.

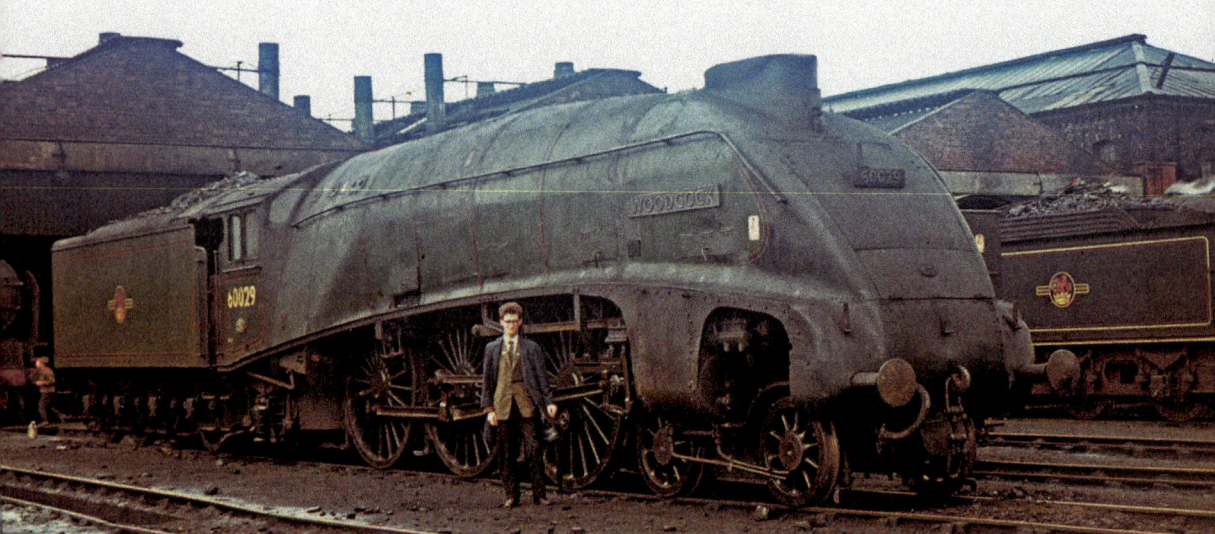

12 May 1963. LNER 4-6-2 A4 No. 60029 Doncaster MPD.

LNER 4-6-2 A4 No. 60029 *Woodcock* is fully coaled and ready for its next duty on the day of the visit by SR WC 4-6-2 No. 34094 *Mortehoe*, which had arrived at Doncaster with a Warwickshire Railway Society special from Birmingham New Street. No. 60029 had been built at the Plant in Doncaster and entered service in July 1937. For most of its career it was a Kings Cross MPD-based locomotive but in June 1963 it was transferred to New England MPD. It had a very short stay there as it had spent nearly the whole of July that year at the Plant undergoing a light casual overhaul. It worked barely three months afterwards before being withdrawn in October 1963.

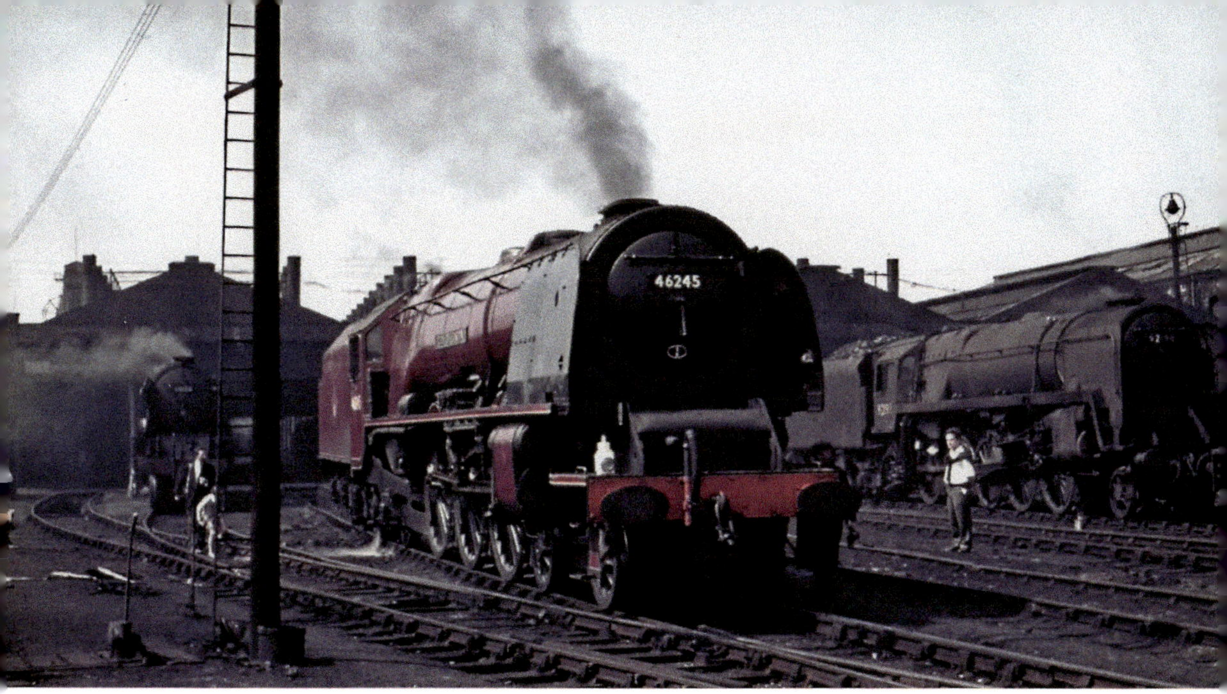

9 June 1963. LMS Duchess 8P 4-6-2 No. 46245 on Doncaster MPD.

LMS 8P 4-6-2 No. 46245 *City of London* hauled a twelve-coach special organised by the Home Counties Railway Society to Doncaster from Kings Cross up the East Coast Main Line on 9 June 1963. The locomotive was still a long-term resident of Camden MPD but its time there was soon to end as it would be transferred to Willesden and finally Crewe North before withdrawal. (L. Flint)

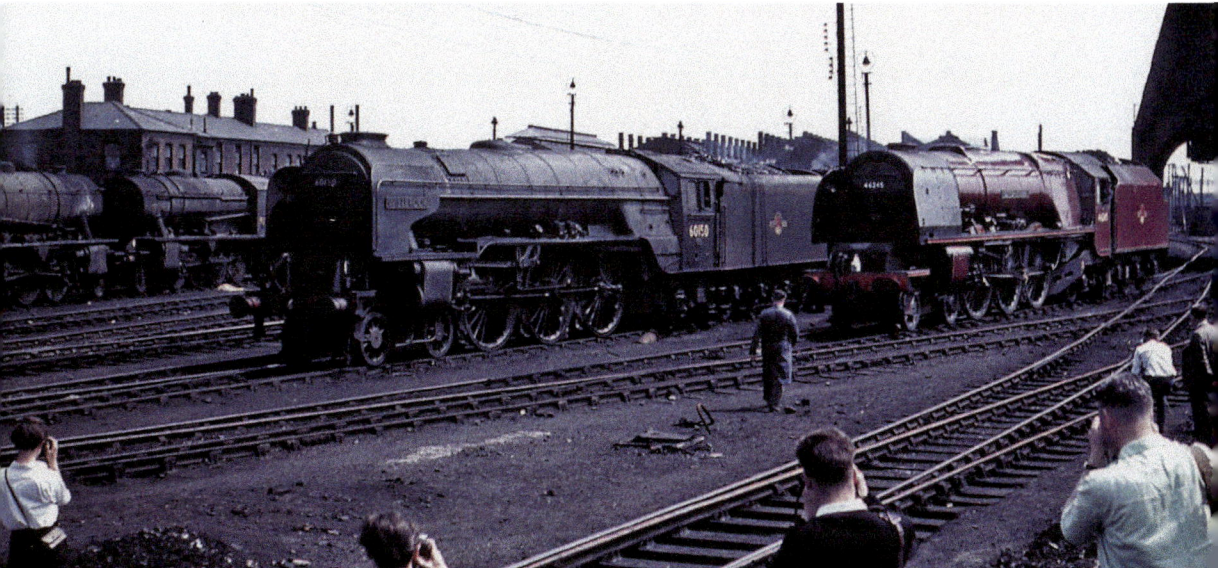

9 June 1963. LMS Duchess 8P 4-6-2 No. 46245 with No. 60150 on Doncaster MPD.

LMS 8P 4-6-2 No. 46245 *City of London* looked quite at home on Doncaster MPD after working the Home Counties Railway Society's Doncaster Special on 9 June 1963. It was seen in the company of LNER A1 4-6-2 No. 60150 *Willbrook* and having been coaled it is on its way to the turntable ready for the return journey to Kings Cross. For both locomotives, their days on the main line were numbered; they would be withdrawn the following year within a month of each other.

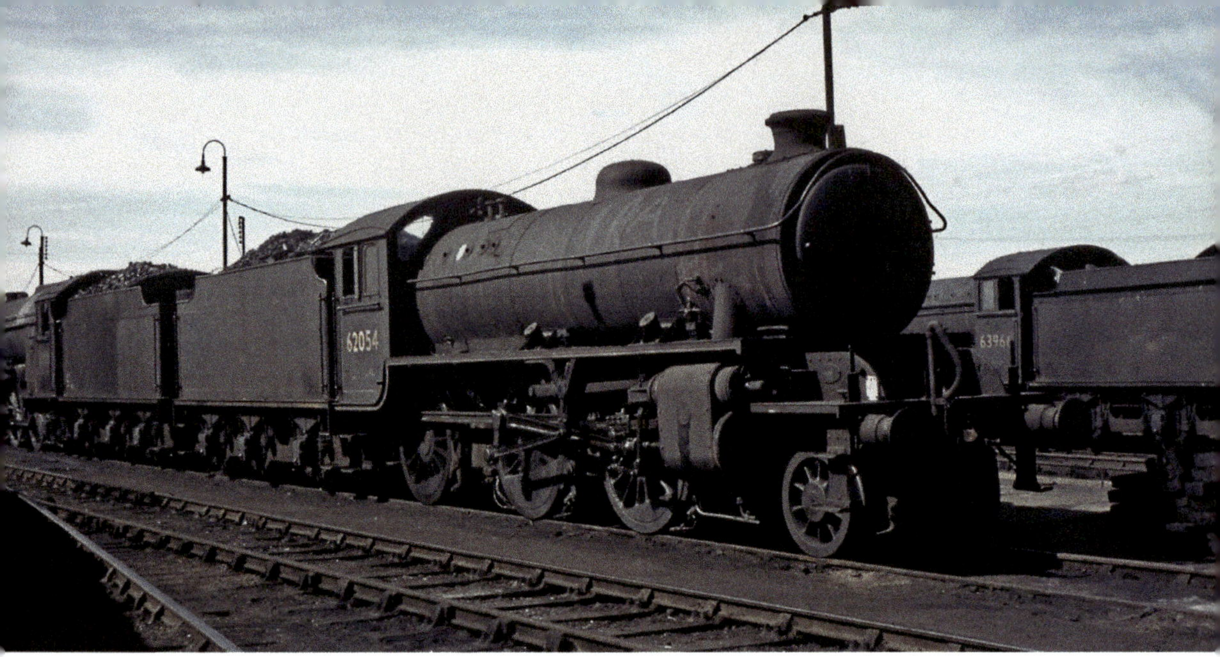

28 July 1963. LNER K1 2-6-0 No. 62054 Doncaster on MPD.

LNER K1 2-6-0 No. 62054 was a product of the North British Locomotive Company of Glasgow and was delivered to the recently formed British Railways in November 1949. Its first allocation was to March MPD, which it served for twelve years before being transferred to Retford Thrumpton MPD for its final three years in service. It was seen on Doncaster MPD on a sunny afternoon eighteen months before its withdrawal.

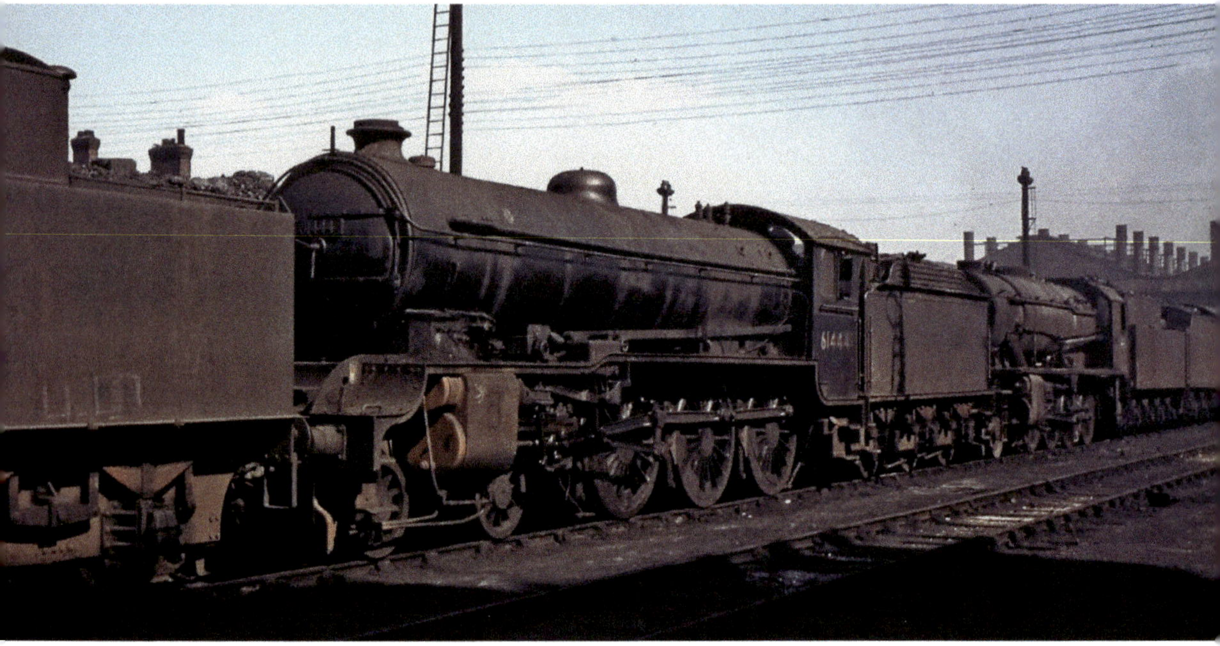

28 July 1963. LNER B16 4-6-0 No. 61444 at Doncaster MPD.

LNER B16/3 4-6-0 No. 61444 had been rebuilt in its final form in 1945 and had been a long-term stalwart of York MPD. The last eighteen months of its service was at Hull's Dairycoates MPD, and it awaits its next duty back there after servicing on Doncaster MPD. It was to see another year of operation before being cut up at Draper's scrapyard in Hull.

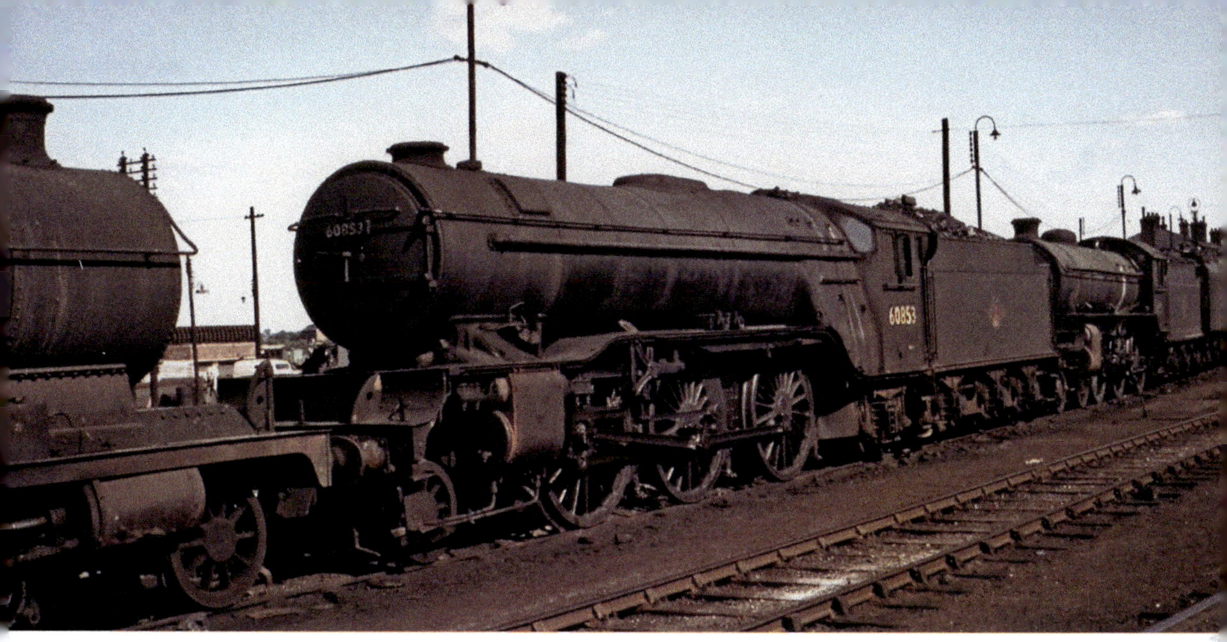

28 July 1963. LNER V2 2-6-2 No. 60853 on Doncaster MPD.

LNER V2 2-6-2 No. 60853 was seen at its home shed, fully coaled and ready for the next duty, but like the other locomotives in the same line it had no duties to perform. It had been built at Darlington in early 1939 and had been allocated to Doncaster for the first six years of its service. In March 1962 it had received a heavy overhaul at the nearby Plant works and was reallocated back to Doncaster MPD. Dieselisation of services on the East Coast Main Line had severely curtailed the use of steam and No. 60853 would be consigned to the scrapyard within a few weeks of this photograph being taken.

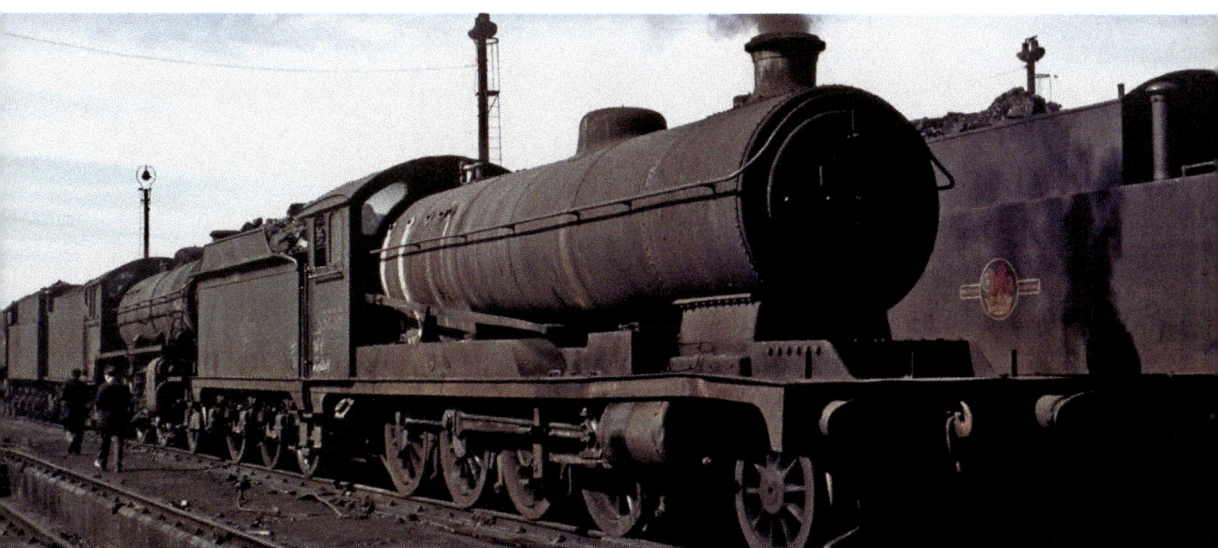

28 July 1963. LNER O4/8 2-8-0 No. 63858 on Doncaster MPD.

LNER 04/8 2-8-0 No. 63858 was a product of the North British Locomotive Company in 1919 and part of an order by the War Department to aid the post-war economy. It had been purchased by the LNER in 1924 and added to their expanding fleet of O4 locomotives. It was reallocated to Doncaster MPD in late 1951 and was to remain faithful to the depot until steam finished there in April 1966. In May 1953 it was rebuilt into its final form as an O4/8 with a BI-type boiler and cab. In common with many O4s it had acquired a very characteristic front-end droop caused by heavy shunting. It is seen in the company of other Doncaster-allocated heavy freight locomotives.

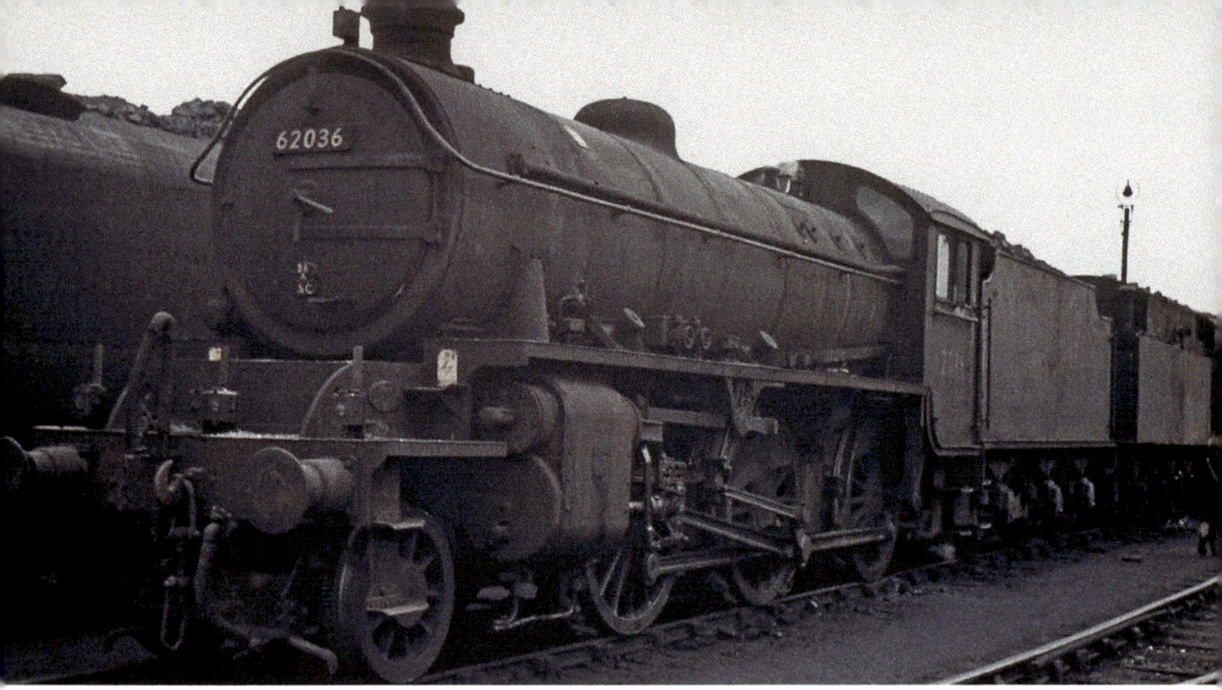

September 1963. LNER K1 2-6-0 No. 62036 at Doncaster MPD.

LNER K1 2-6-0 No. 62036 had been allocated to Doncaster MPD since January 1961 but was within weeks of withdrawal when it was photographed on its home shed. It had been built to an LNER design by the North British Locomotive Company, Glasgow, and delivered to British Railways in September 1949. It completed just fourteen years of service before it was declared surplus to requirements.

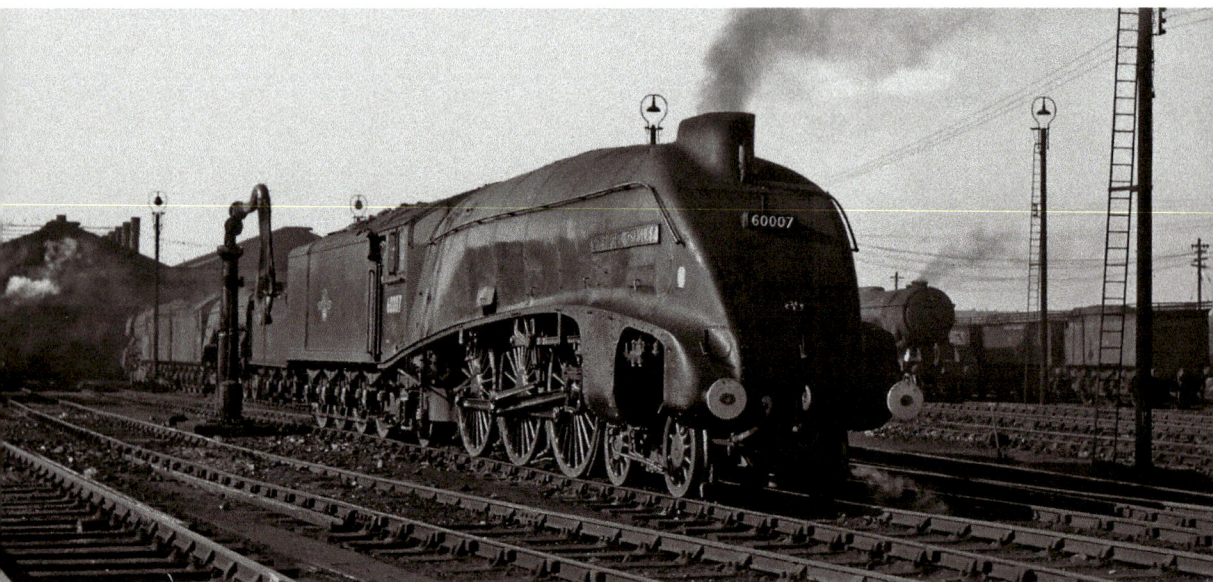

8 September 1963. LNER A4 4-6-2 No. 60007 on Doncaster MPD.

LNER A4 4-6-2 No. 60007 *Sir Nigel Gresley* had been a Kings Cross locomotive for much of its life but by 1963 much had changed on the East Coast Main Line and No. 60007 had found a temporary home at New England MPD. Within a month of this photograph being taken the loco, along with other A4s, would be transferred to Scotland for a few years of express duties. When it was finally withdrawn in February 1966, *Sir Nigel Gresley* was bought for preservation. Since then it has regularly worked special trains over the national network. (L. Flint)

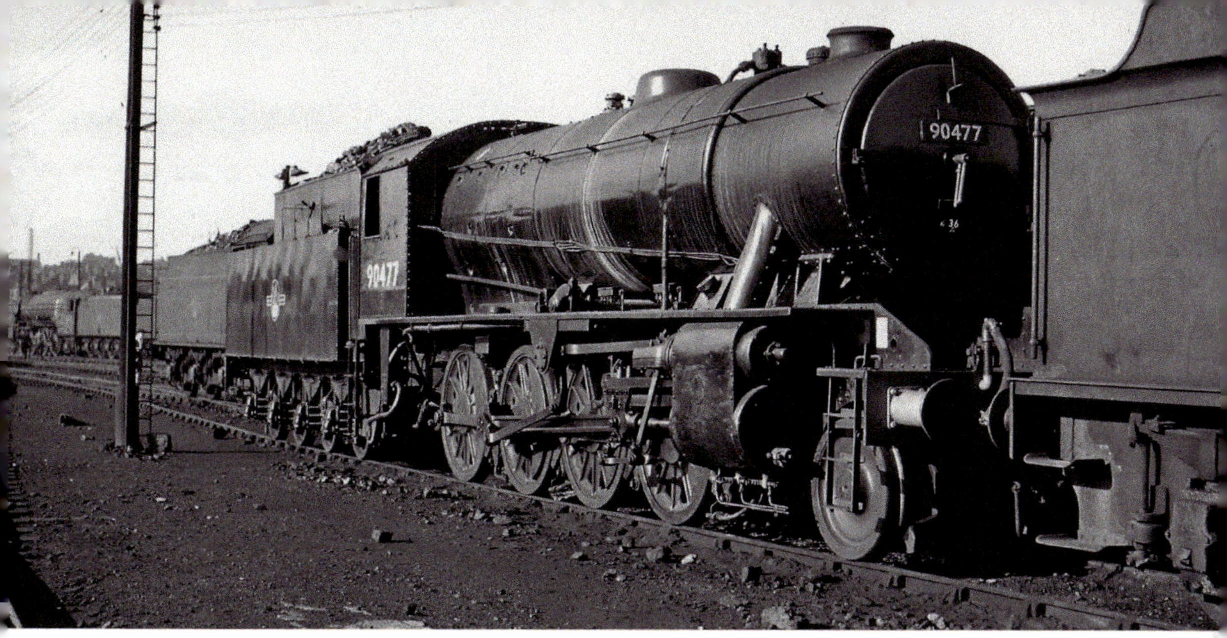

8 September 1963. WD 2-8-0 No. 90477 on Doncaster MPD.

WD 2-8-0 No. 90477 had been a Doncaster-allocated locomotive from October 1961 until its withdrawal in March 1966, apart from a ten-week stay at Immingham MPD. It was seen here on its home depot having just returned to service after its final general overhaul at the Plant. WDs in this condition were an unusual sight because they rarely received any attention from the cleaners. Even though No. 90477 glinted in the afternoon sunshine, it had already begun to collect limescale stains on its boiler. With a huge supply of good quality coal in its tender it looked ready for its next task. (L. Flint)

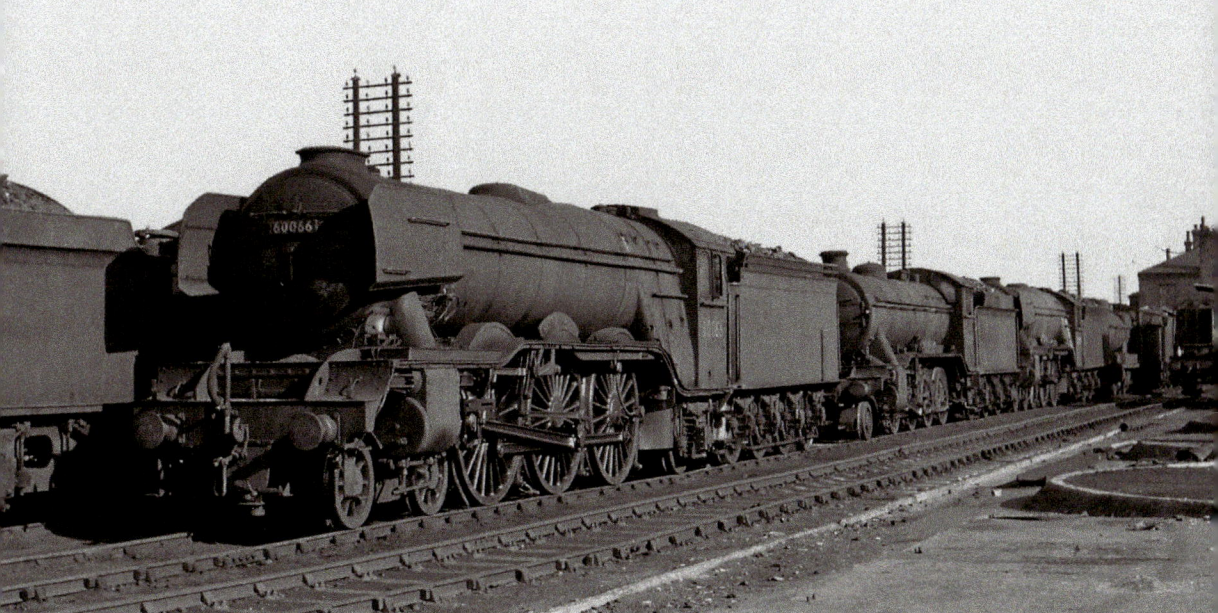

8 September 1963. LNER A3 4-6-2 No. 60066 on Doncaster MPD.

LNER A3 4-6-2 No. 60066 *Merry Hampton* stands on Doncaster MPD on the day it was officially withdrawn after thirty-nine years of service on the East Coast Main Line. Although looking rather unkempt and with scorched smokebox door, its pedigree still shines through. (L. Flint)

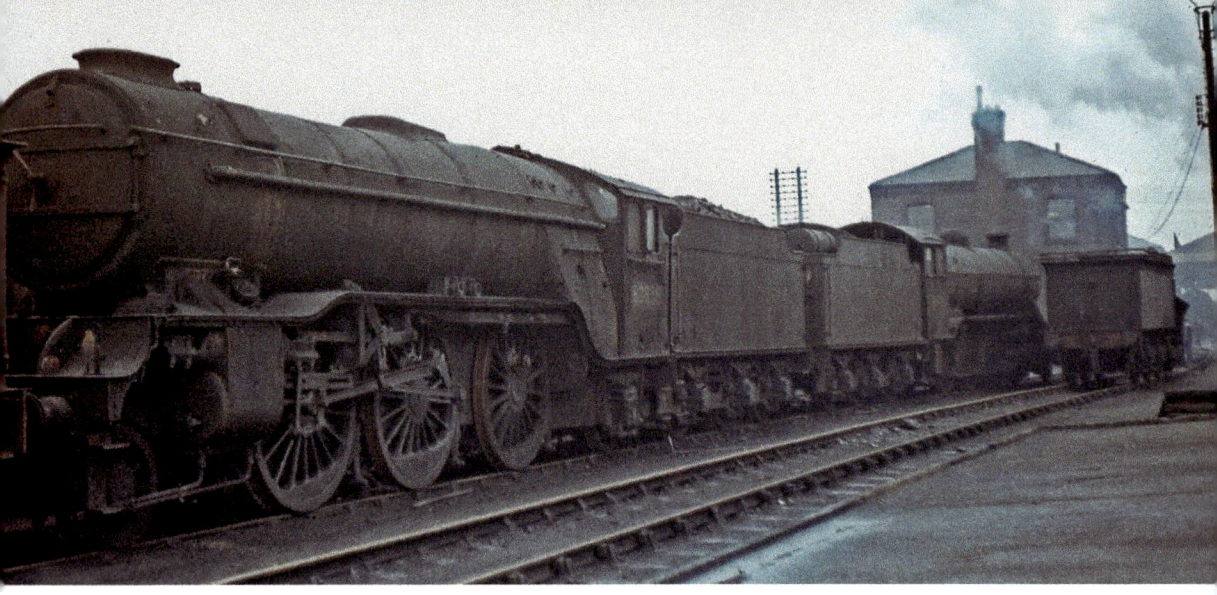

2 January 1964. LNER V2 2-6-2 No. 60858 on Doncaster MPD.

LNER V2 2-6-2 No. 60858 was one of the six V2s that had been modified with a double chimney and Kylchap cowl in 1961. It was said that this late modification gave these locomotives equal performance to A3 Pacifics. Very quickly their primary work had been taken by diesels however, and modification of further class members was abandoned. No. 60858 had been a New England locomotive since 1961 and is in typical condition of engines from that depot. It was reallocated to Doncaster MPD for the last four months before its withdrawal in October 1963. Within days of it being photographed it was hauled across to the Plant for cutting up.

April 1964. LNER O4/8 2-8-0 No. 63613 on Doncaster MPD.

LNER O4/8 2-8-0 No. 63613 had been built by the Great Central Railway in February 1914 and rebuilt as an O4/8 by the LNER in August 1944. This modification involved replacing the original boiler and cab while retaining the wheels, frames, cylinders and tender of the original locomotive. No. 63613 was one of the first to undergo this type of rebuild and in total ninety-nine locos were rebuilt, the last in 1958.

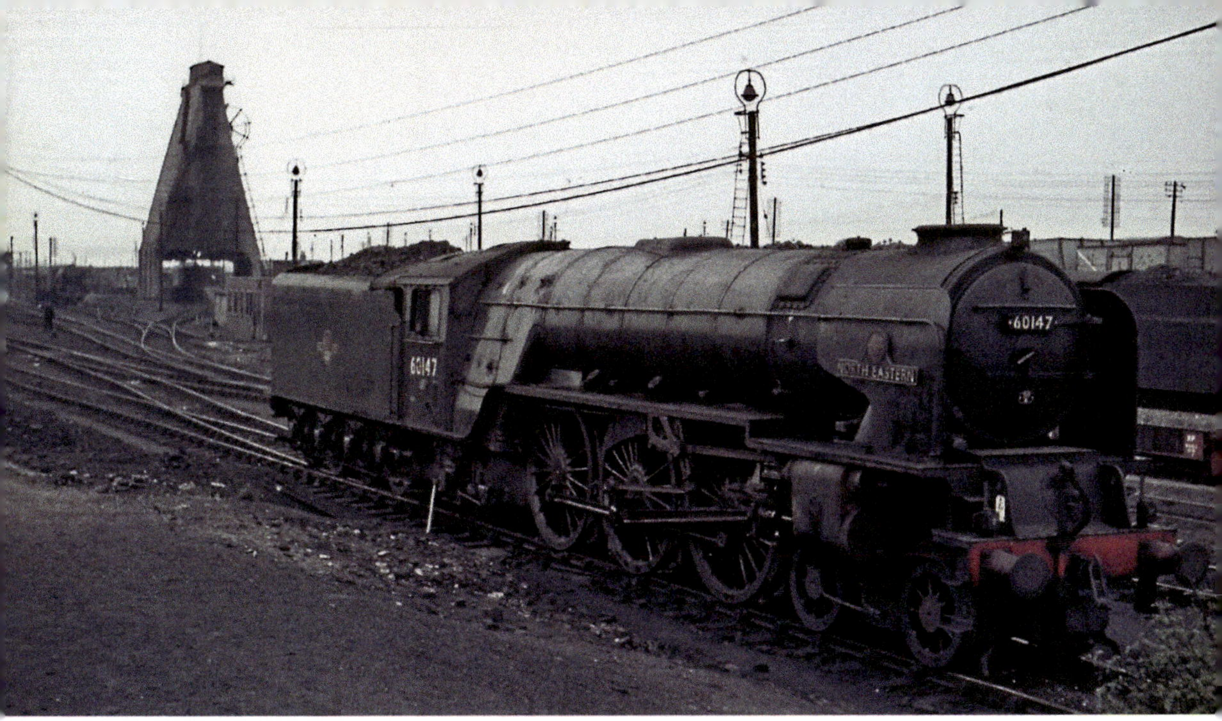

26 April 1964. LNER A1 4-6-2 No. 60147 on Doncaster MPD.

LNER A1 4-6-2 No. 60147 *North Eastern* was another of Peppercorn's fine Pacifics to find sanctuary at York MPD. It would be a short reprieve for the locomotive and despite its reasonable outward appearance in this image it would be withdrawn before the end of August in 1964. (L. Flint)

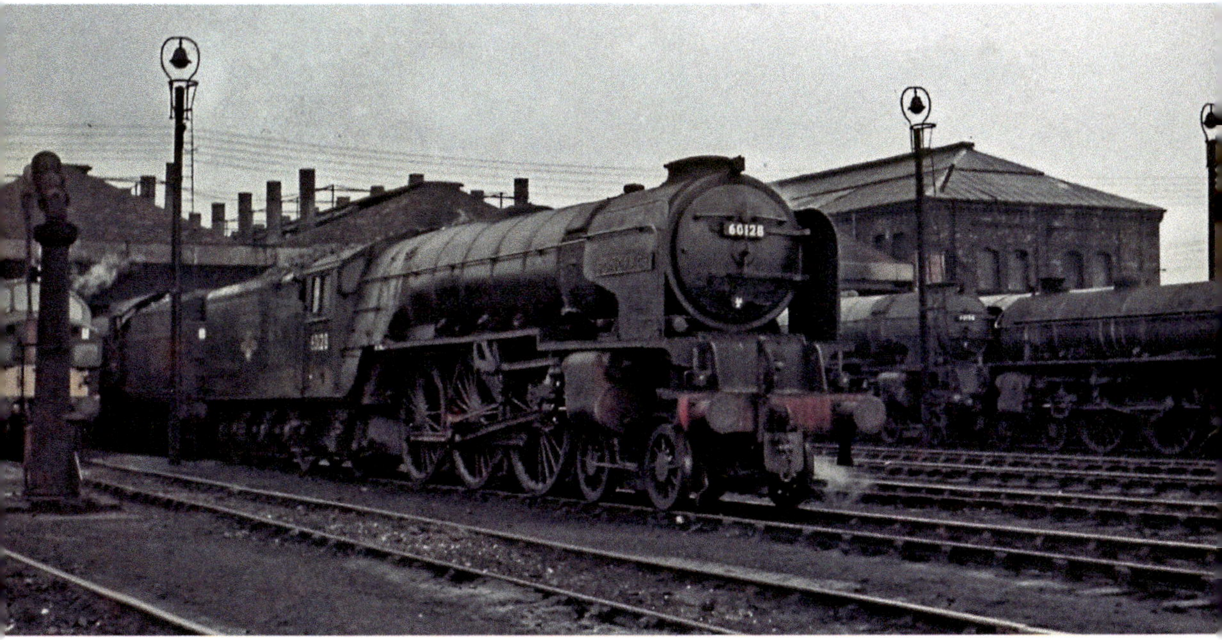

26 April 1964. LNER A1 4-6-2 No. 60128 on Doncaster MPD.

LNER A1 4-6-2 No. 60128 *Bongrace* had been a Doncaster-based locomotive since April 1959. It was a familiar sight during visits to Doncaster and was regularly used as the standing pilot at the south end of the station. It would remain in active service into 1965 before being withdrawn in January of that year. (L. Flint)

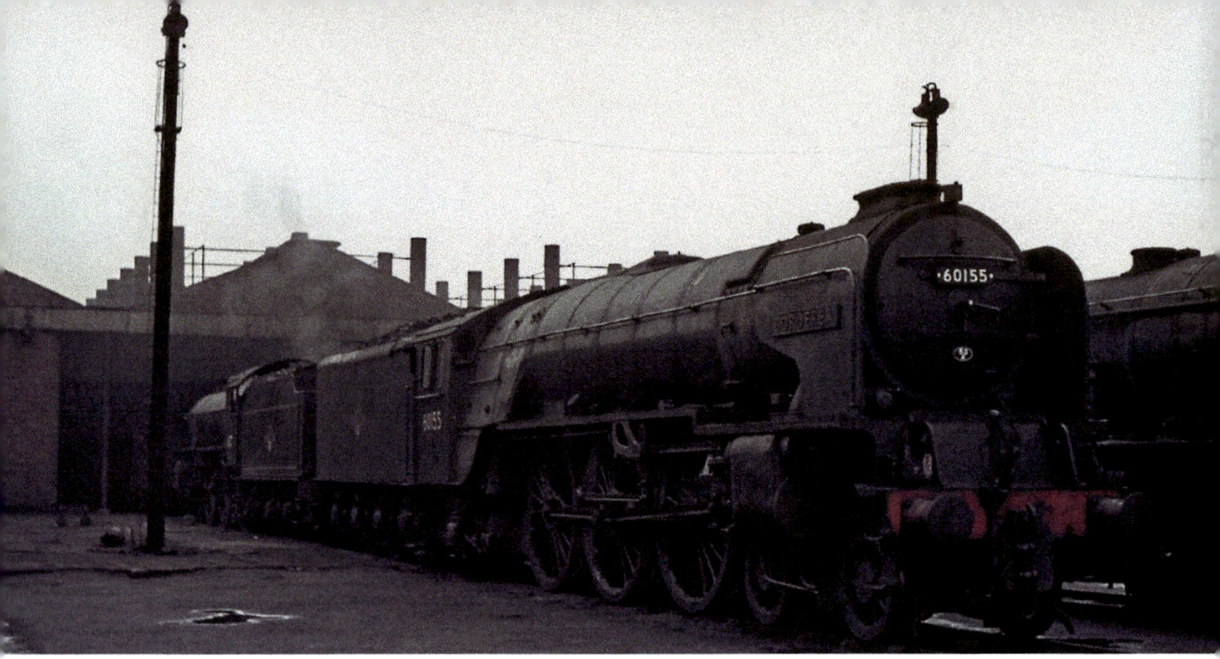

26 April 1964. LNER A1 4-6-2 No. 60155 on Doncaster MPD.

LNER A1 4-6-2 No. 60155 *Borderer* had been a resident of York MPD from late 1962 and remained there until its withdrawal from service in October 1965. Its days working top link East Coast Main Line expresses had by 1964 become a memory as it had worked into Doncaster earlier in the day on a York local train and would take the return working home later. (L. Flint)

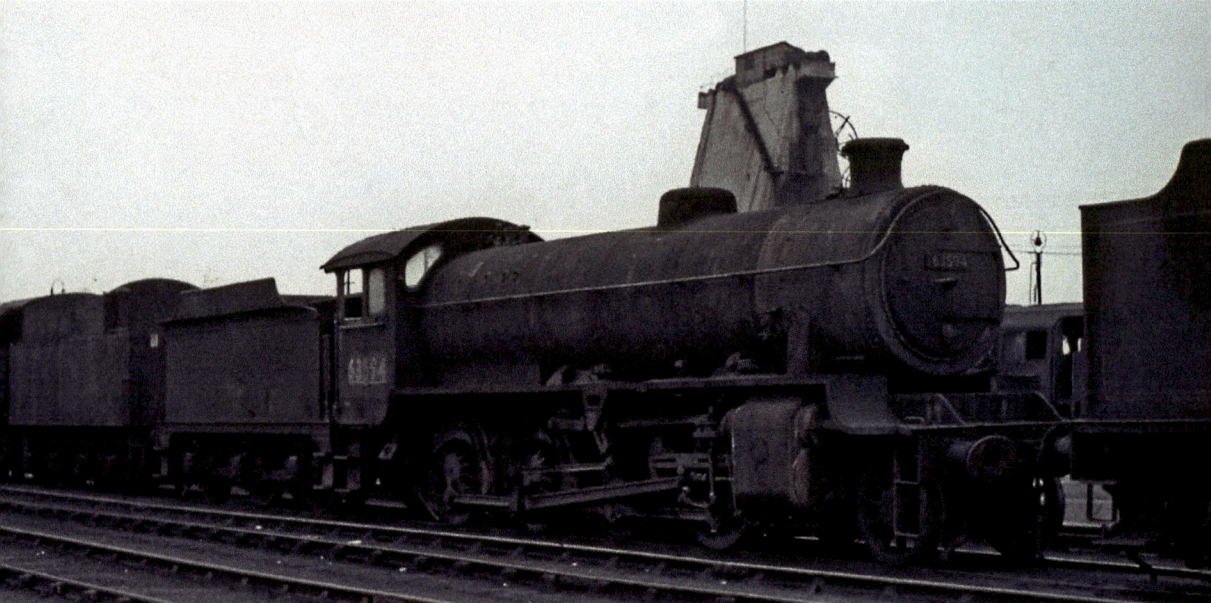

26 April 1964. LNER O1 2-8-0 No. 63594 on Doncaster MPD.

LNER O1 2-8-0 No. 63594 had originally been built as an O4/1 in 1911 and had been rebuilt in 1947 with new cylinders, boiler and cab as an O1. It had spent most of its long career working from depots connected to the Great Central Railway and had been withdrawn a few days before being photographed on Doncaster MPD. It was to remain here with other withdrawn locomotives until being towed away to Smith's yard at Ecclesfield to help feed the blast furnaces of Rotherham and Sheffield. (L. Flint)

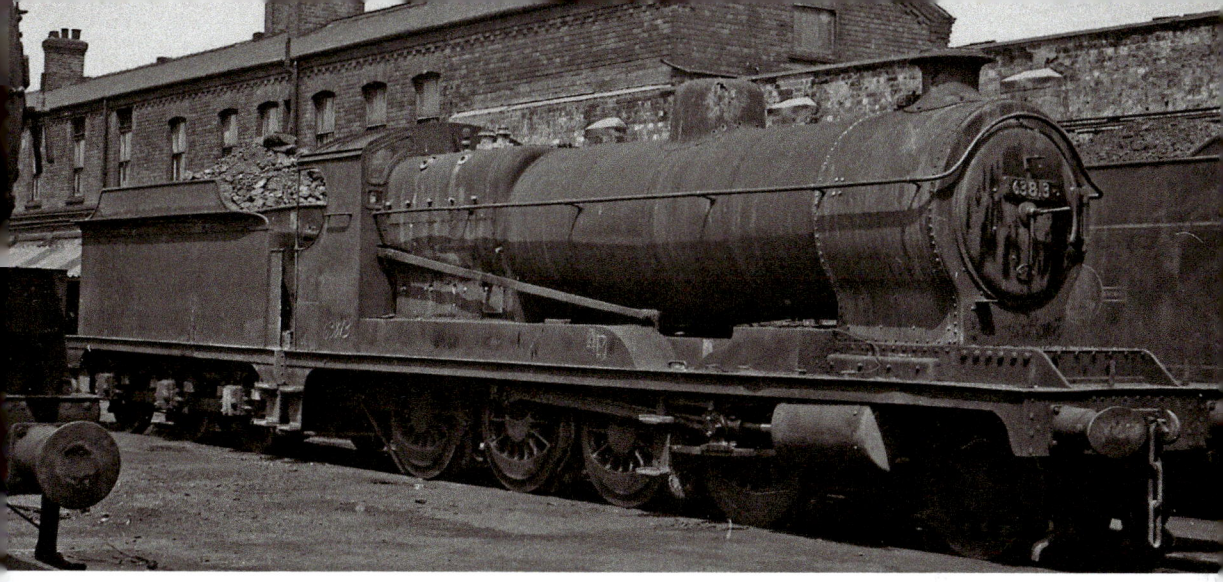

30 May 1964. LNER O4 2-8-0 No. 63813 on Doncaster MPD.

LNER O4 2-8-0 No. 63813 had been built for the Railway Operating Division of the Royal Engineers by the North British Locomotive Company of Glasgow to the original Great Central Railway design in 1918. It was one of the 521 such locomotives built under this order. Over 300 were to work in Europe but the remainder were to be stored immediately after building. Eventually the locomotives were sold off by the government over a number of years, with this locomotive being part of the last batch of a hundred sold to the LNER in 1928 for the princely sum of £340 per loco. Although looking very careworn in this image and with rust holes in its dome cover, it would continue for another ten months before withdrawal. Over thirty-five years of continuous service must represent the best value for money of any locomotive.

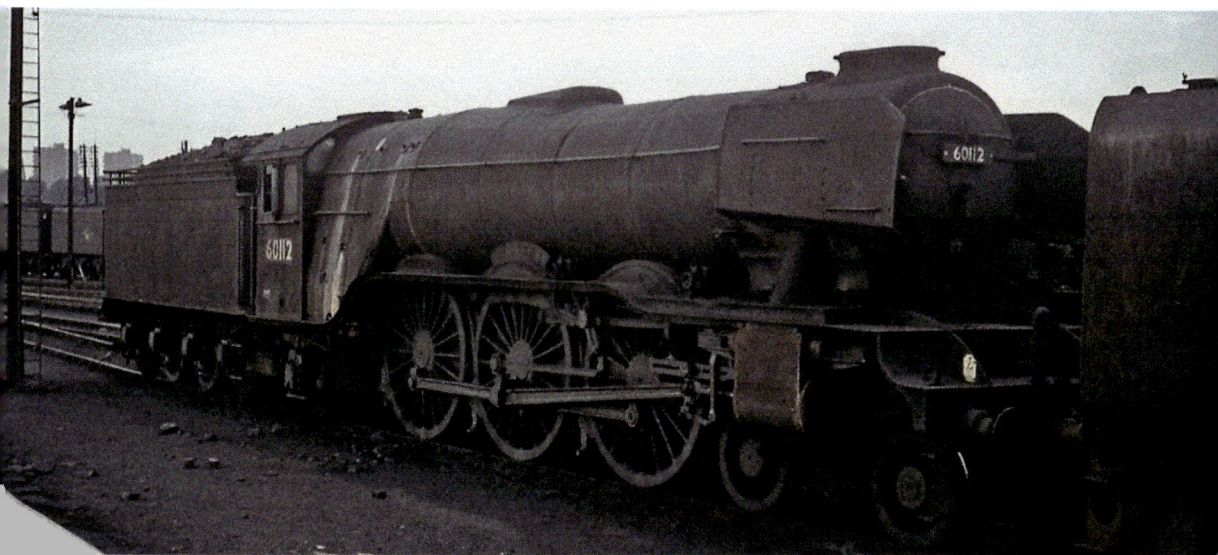

14 July 1964. LNER A3 4-6-2 No. 60112 on Doncaster MPD.

LNER A3 4-6-2 No. 60112 *St Simon* had many connections with Doncaster in its long career. It had been built at the Plant in August 1923 as an A1 and had been allocated to the motive power depot on nine different occasions. In common with many A3 class locomotives, No. 60112 had been named after a racehorse; St Simon had won the Princess of Wales's Nursery Plate at Doncaster and was ranked as the fourth best English racehorse of the nineteenth century by racing experts of that time. No. 60112 had been photographed here near the end of its working life and it was withdrawn six months later.

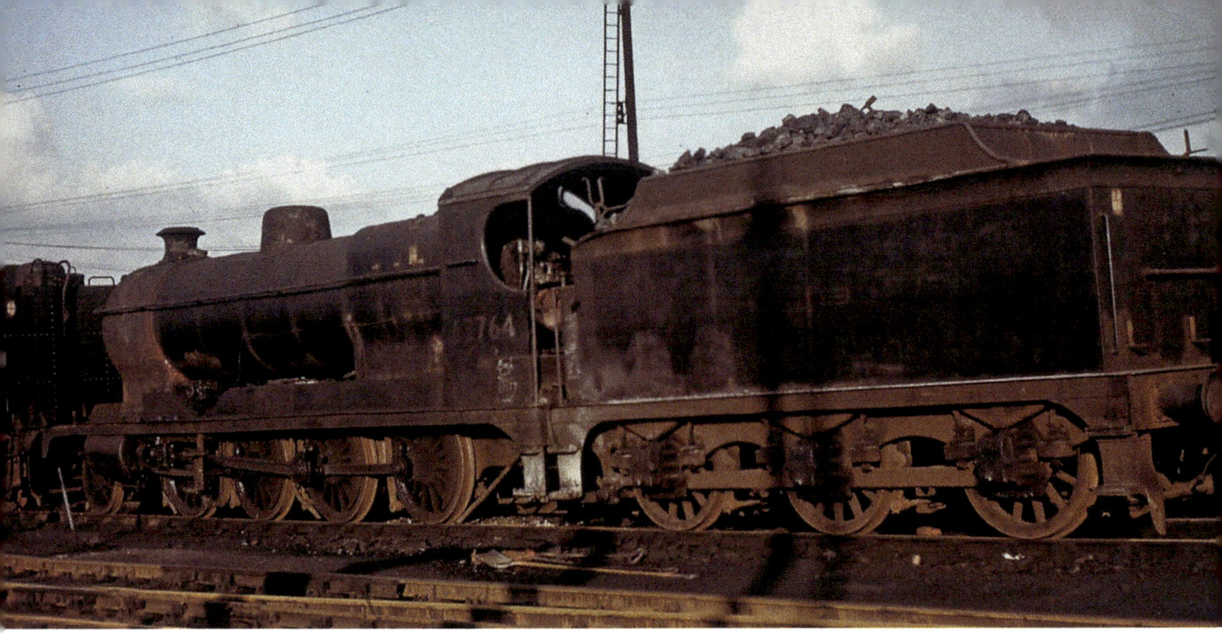

September 1964. LNER O4 2-8-0 No. 63764 on Doncaster MPD.

LNER O4 2-8-0 No. 63784 was another of the many Great Central Railway-designed locomotives built for the Railway Operating Division of the Royal Engineers by the North British Locomotive Company of Glasgow. It was completed in 1918, too late for service abroad, and was loaned out to the LNWR for a short time. It was one of the hundred locomotives in the last batch sold to the LNER in 1927. Although it appears to have rusty wheel faces in this image, it does have a good load of decent coal and would remain in service for another eighteen months before its final withdrawal.

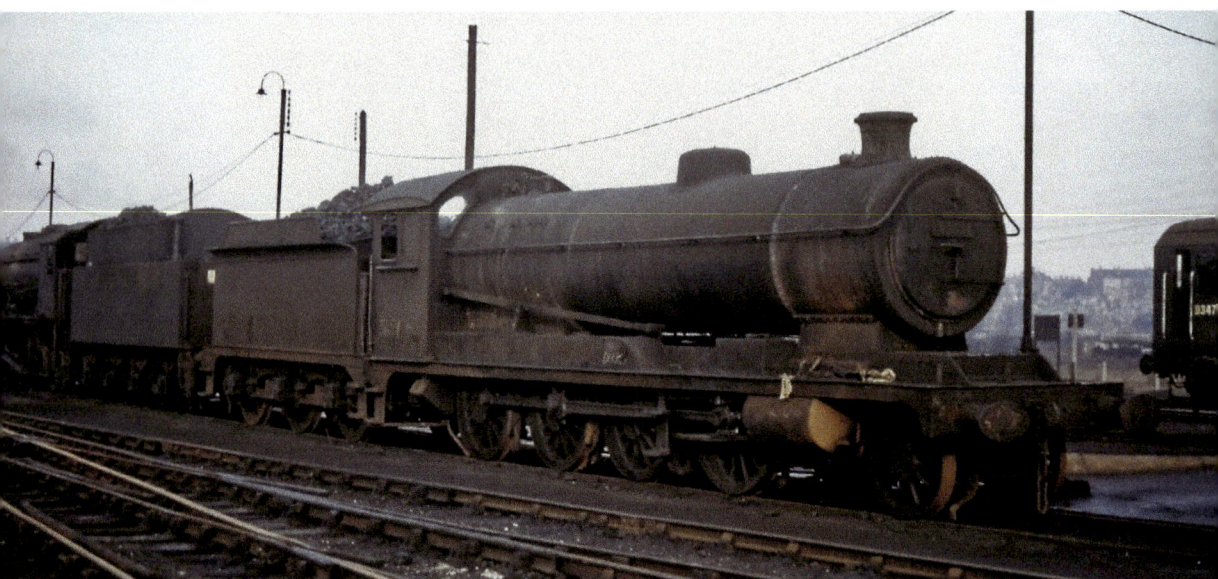

October 1964. LNER O4/8 2-8-0 No. 63651 on Doncaster MPD.

LNER O4/8 2-8-0 No. 63651 had been built by Kitson & Company of Hunslet, Leeds, as part of an order of thirty-two locomotives for the Ministry of Munitions in 1918. It was eventually bought by the LNER in early 1925, one of a batch of forty-eight acquired that year. In December 1946 it was rebuilt at Gorton Works as an O4/8 with a B1 type boiler and cab. Although looking in poor external condition at the time of the photograph, the Frodingham-based loco would continue in service for another ten months.

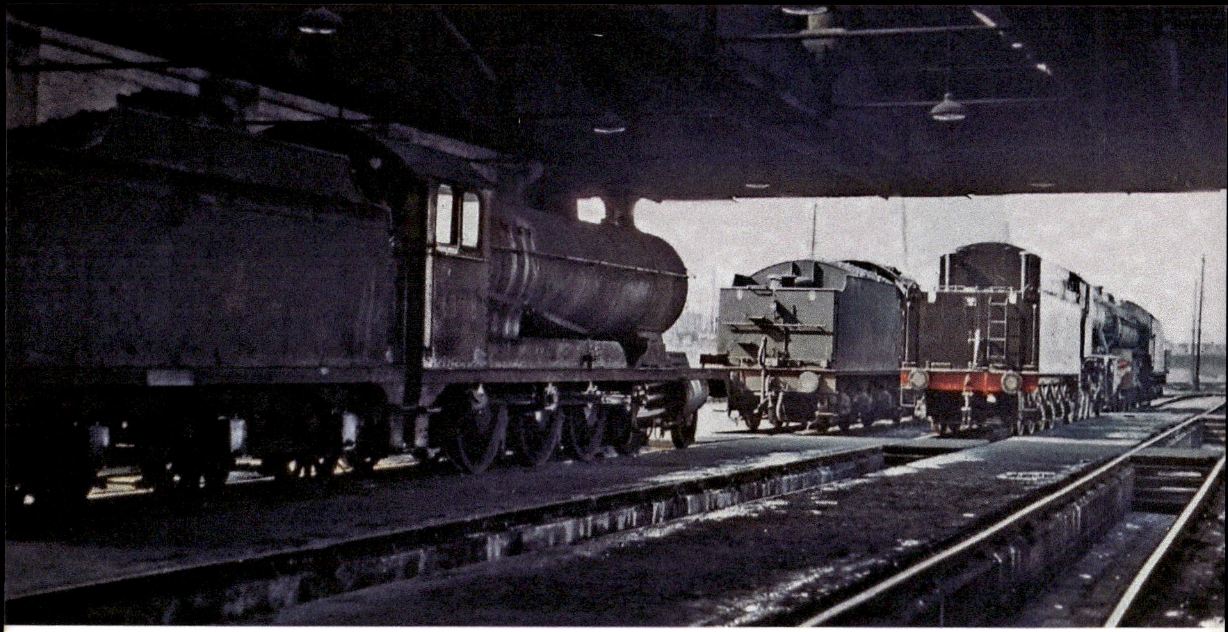

March 1965. LNER O4/8 2-8-0 No. 63818 inside Doncaster MPD.

LNER O4/8 2-8-0 No. 63818 shelters inside its home shed of Doncaster. On the adjacent road a WD 2-8-0 stands resplendent in the sunshine after returning from the Plant with a new coat of paint after its last general overhaul. By this time the depot was inhabited by mainly heavy freight locomotives. The graceful Gresley and Peppercorn Pacifics, which had for so long been a familiar sight, were now all gone and within a year the depot would close to all steam.

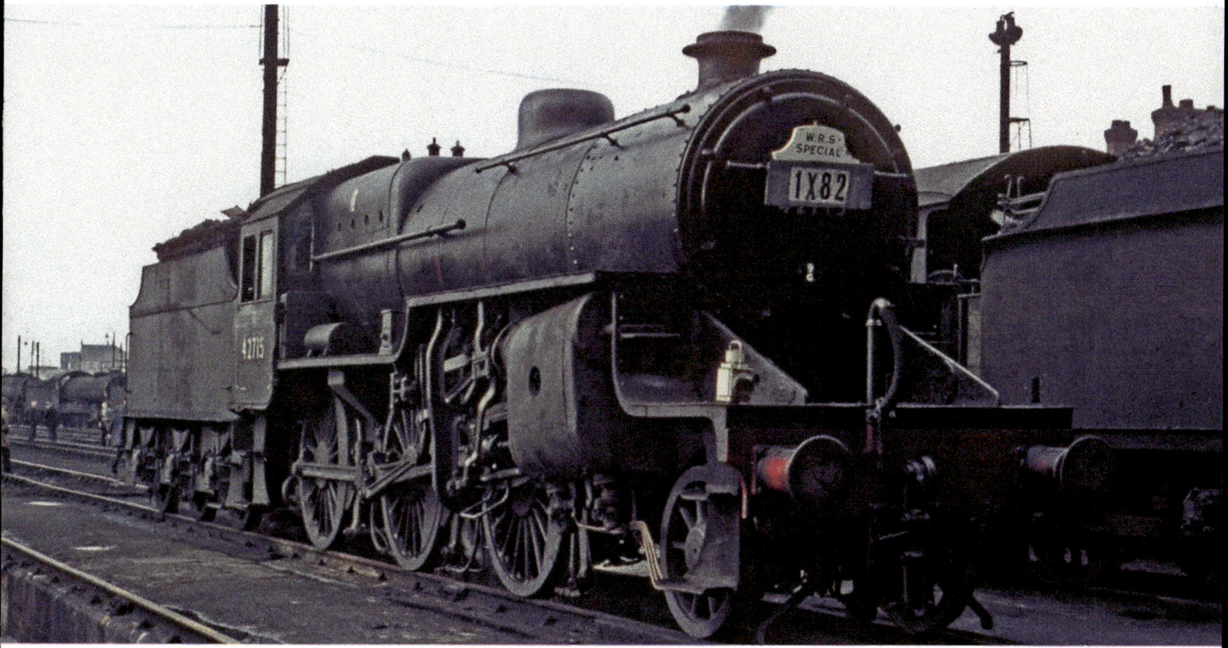

8 May 1965. LMS Crab 2-6-0 No. 42715 at Doncaster MPD.

LMS Crab 2-6-0 No. 42715 serviced on Doncaster MPD. The locomotive had worked in on the Warwickshire Railway Society South Yorkshire and Nottinghamshire Rail tour, which it had taken over at Mansfield and would return it to later in the day. No. 42715 was a Gorton-allocated loco at this time and had been built in 1927. It was eventually withdrawn in February 1966. It was quite unusual to see this class of locomotive in Doncaster, although they did appear occasionally working through on seaside excursion trains.

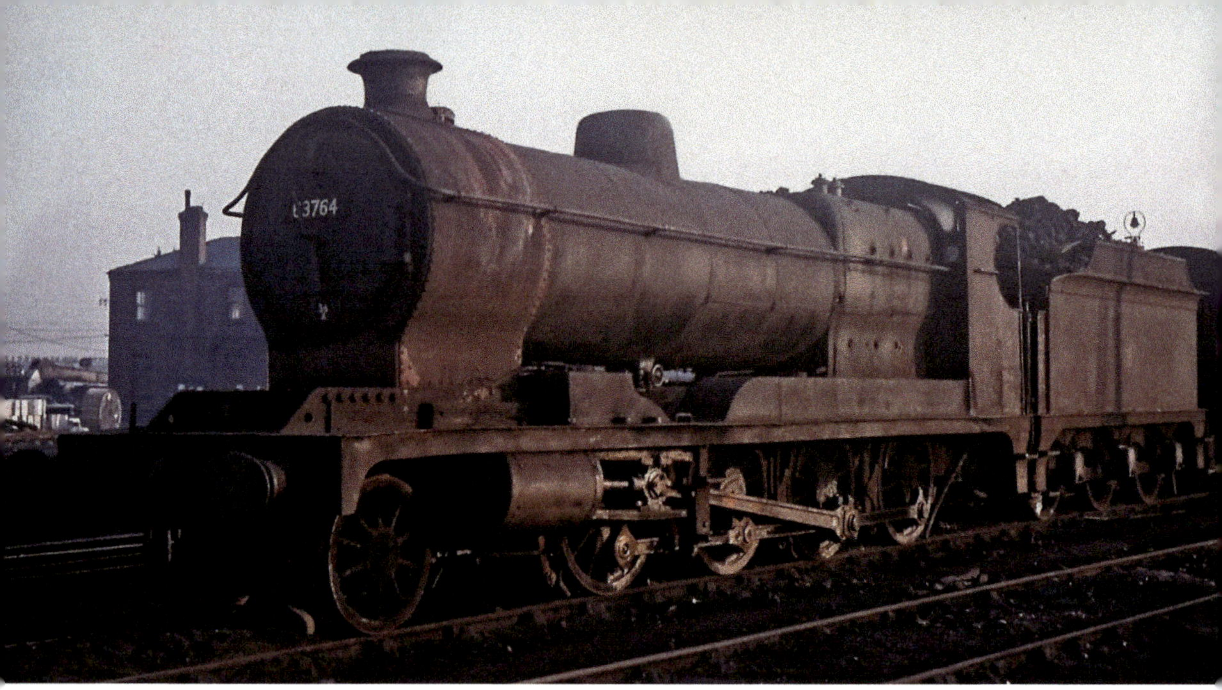

27 December 1965. LNER O4 2-8-0 No. 63764 on Doncaster MPD.

LNER O4 2-8-0 No. 63764 at Doncaster MPD was fully coaled and ready for work as it waited in the winter sunshine after the Christmas break. The new year was not to be so happy for the forty-eight-year-old veteran though, as it was withdrawn at the end of February 1966. It was scrapped by T. W. Ward at the Beighton yard a few months later.

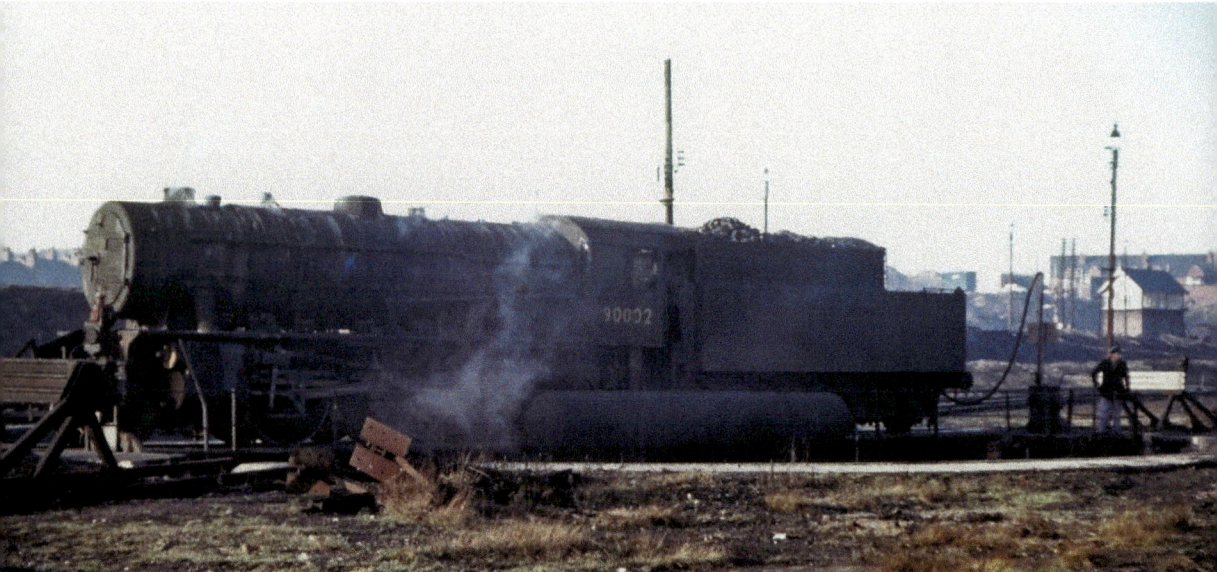

January 1966. WD 2-8-0 No. 90002 on Doncaster MPD.

WD 2-8-0 No. 90002 had been coaled and was being turned on the Doncaster shed turntable as preparation for its next duty. This locomotive was one of a number to arrive at Doncaster in January 1966 from Colwick MPD when that depot had been transferred to LMS control. The new regime there moved LMS locomotives in to take over the duties. No. 90002 had only a short stay at Doncaster because it was withdrawn in April 1966 and broken up by Draper's of Hull in that summer.

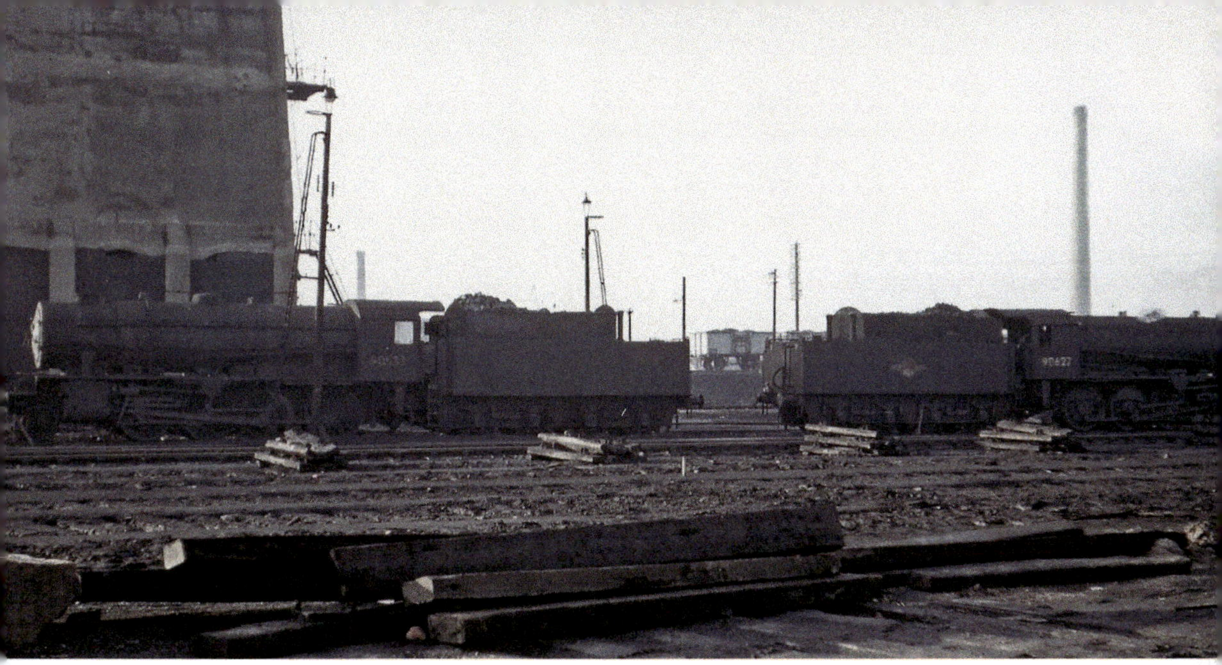

January 1966. WD 2-8-0s Nos 90533 and 90627 on Doncaster MPD.

WD 2-8-0s Nos 90533 and 90627 were seen on Doncaster MPD in early 1966. No. 90533, on the left, was another exile from Colwick MPD. It was to find little favour at Doncaster and was withdrawn very soon after this photograph was taken. No. 90627, the one on the right, faired a little better. It was a long-term resident of Hull Dairycoates MPD and remained there until the depot closed in June 1967. It moved to West Hartlepool to see out the end of steam there before withdrawal and scrapping by Hughes Bolckow at North Blyth.

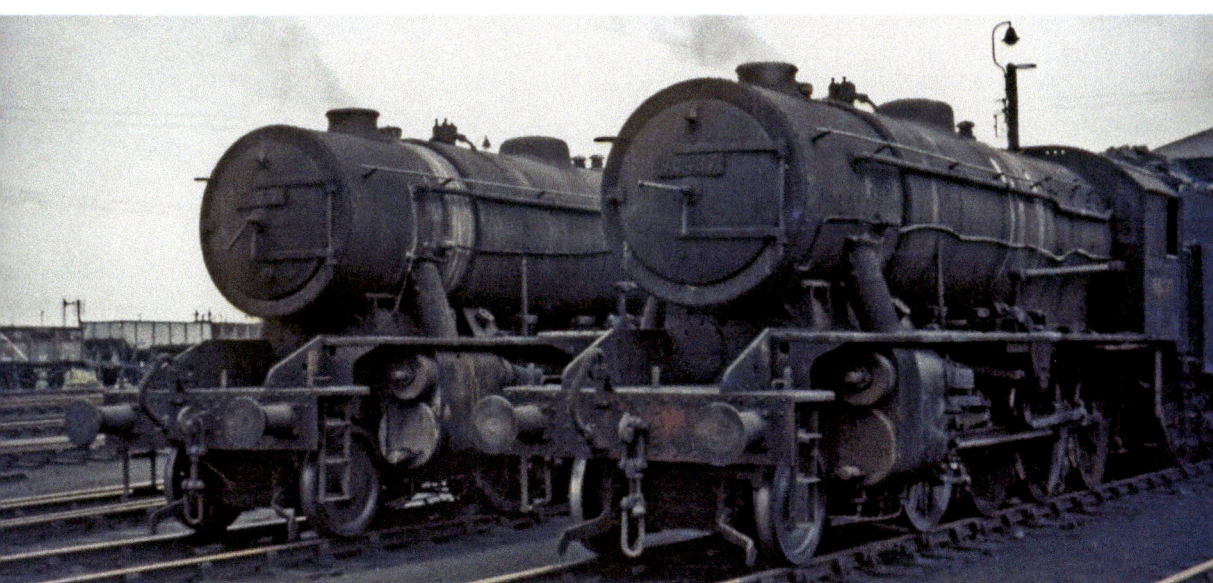

February 1966. WD 2-8-0s Nos 90037 and 90437 on Doncaster MPD.

WD 2-8-0s Nos 90037 and 90437 are prepared at Doncaster MPD for more work in the coalfields. It was towards the very end of steam at Doncaster and these two were weeks away from withdrawal. Although there were 400 locomotives between them numerically, these locos were completed within weeks of each other in early 1944. No. 90037 had been built by the North British Locomotive Company of Glasgow while No. 90437 came from the Vulcan Foundry at Newton-le-Willows.

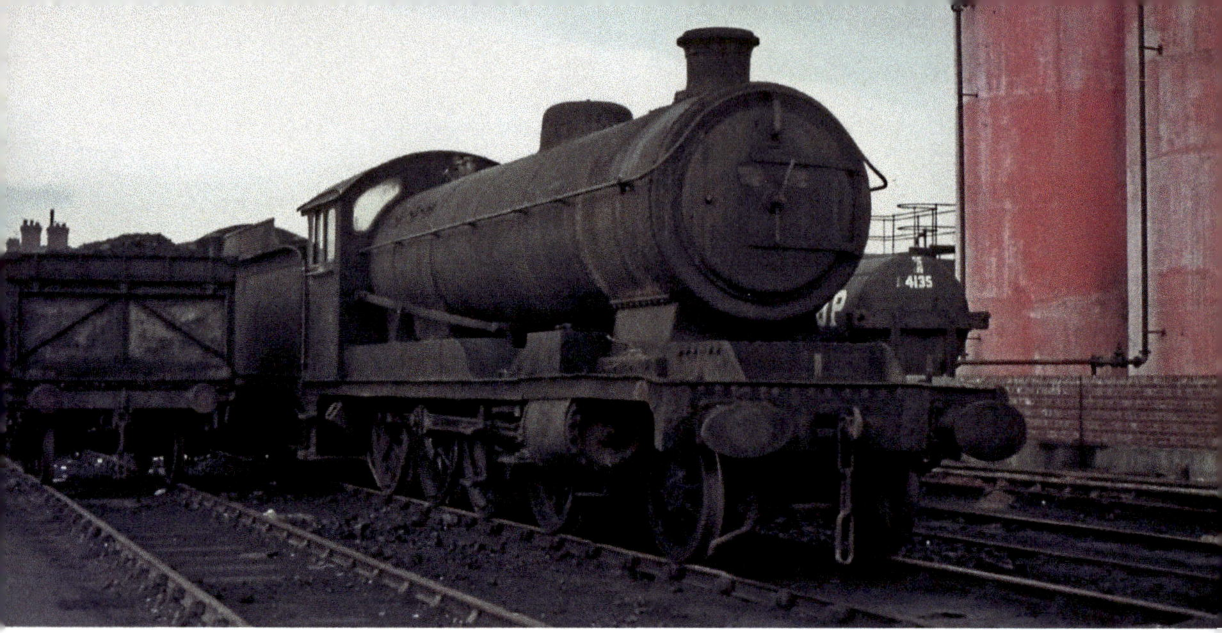

30 May 1966. LNER O4/8 No. 63818 on Doncaster MPD.

LNER O4/8 No. 63818 was photographed on Doncaster MPD a month after it had been withdrawn from service and awaited its last journey to W. George of Station Steel, based at the closed Wath Central station. No. 63818 had been constructed in July 1918 by the North British Locomotive Company of Glasgow as part of an order for 344 similar locomotives from the Ministry of Munitions for service in France during the First World War. It was part of the first batch of O4s purchased by the LNER in late 1924. In June 1947 it was chosen as one of ninety-nine O4s to be rebuilt with B1 type boiler and cab as O4/8s. (L. Flint)

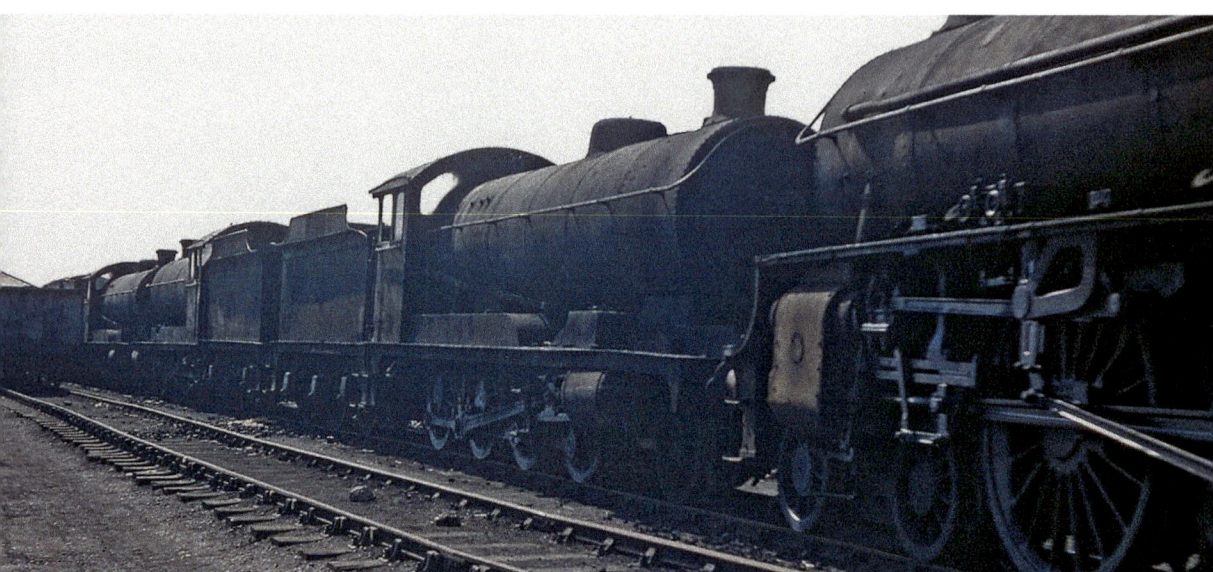

May 1966. LNER O4/8 2-8-0s at Doncaster MPD.

LNER O4/8 2-8-0s Nos 63781, 63653 and 63868 keep company with an LNER B1 4-6-0 as they await their removal to the scrapyard. In the case of the O4s, it was the yard of W. George & Son, known as Station Steel of Wath-on-Dearne, which was situated next to Wath Central station. Doncaster MPD had closed to steam a few weeks before this image and the sad lines of redundant locomotives, which had been the shed's final allocation, would soon disappear.

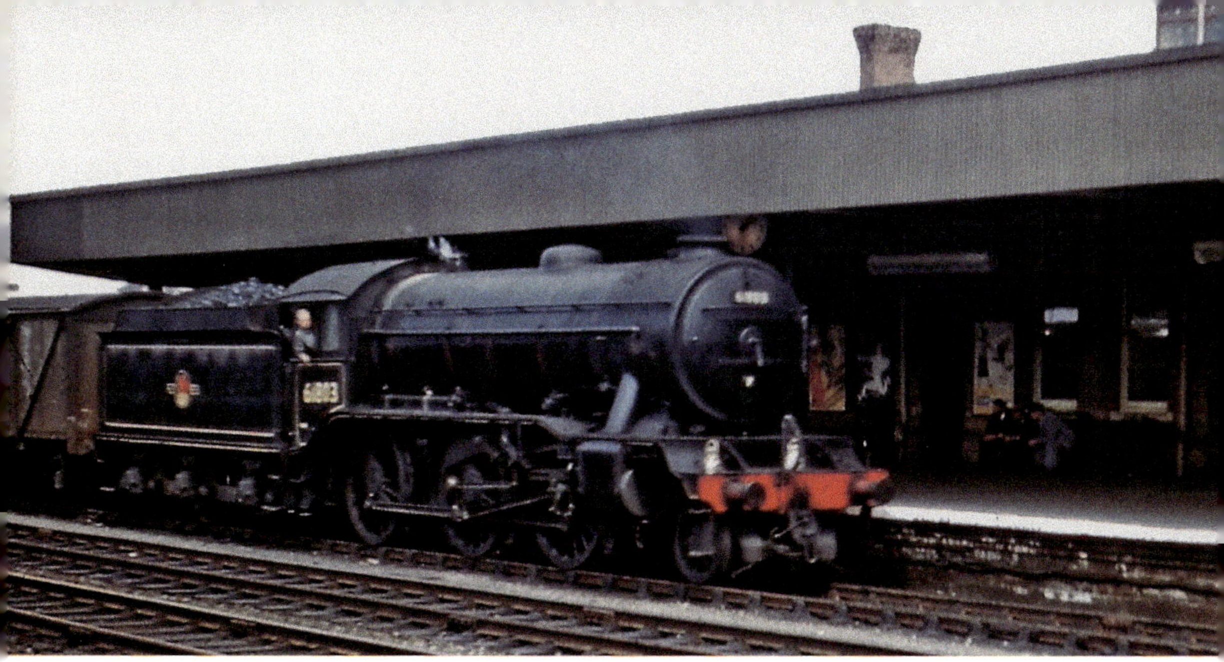

May 1960. LNER K3 2-6-0 No. 61803 heads through Doncaster station.

LNER 2-6-0 K3 No. 61803 heads a Grimsby fish train through Doncaster station heading towards Sheffield and leaving the distinctive smell lingering on the platforms for a while after it had passed through. It would have been a familiar working for the locomotive, which had spent much of its time allocated to Immingham shed. Looking in good fettle, it seems surprising that the loco had just over a year left in service. In fact, the whole class of 193 locomotives had been withdrawn from active service by December 1962.

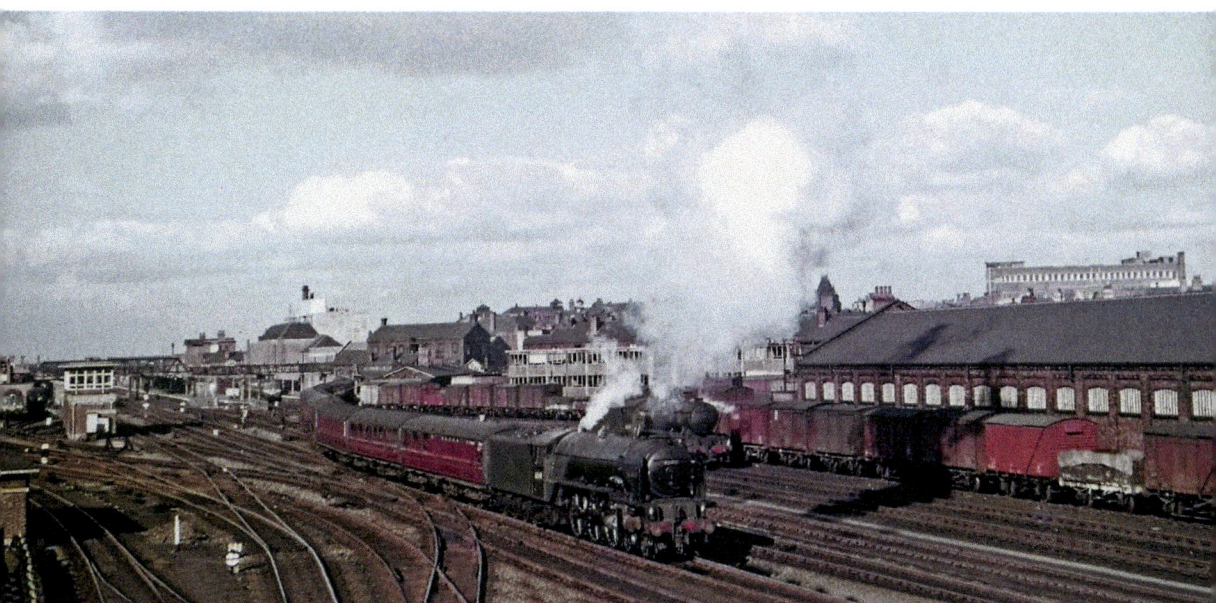

June 1960. LNER A1 4-6-2 No. 60149 departs from Doncaster station.

A typical scene from St James' Bridge looking north towards the station with LNER 4-6-2 A1 No. 60149 *Amadis*, a Doncaster engine, departing south from Doncaster station's platform 1. It is passing a V2 in the usual position of the southbound standby loco. The bridge area was a prime spot for trainspotters after they had been moved off the station platforms by the authorities.

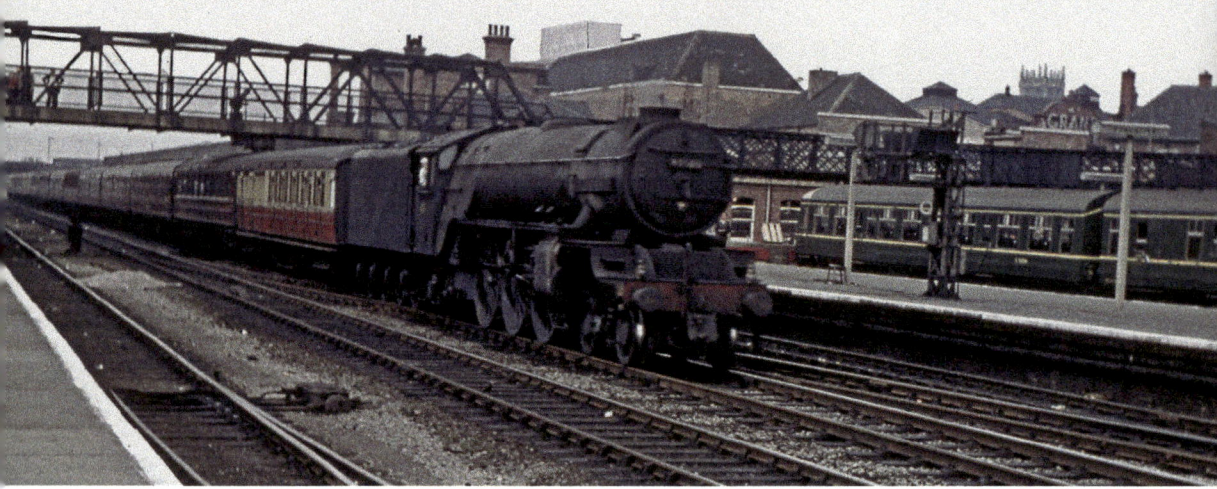

June 1960. LNER A2/2 4-6-2 No. 60506 speeds through Doncaster station.

LNER A2/2 4-6-2 No. 60506 *Wolf of Badenoch* rushes through Doncaster station with a southbound express. No. 60506 was one of a class of six P2/2 2-8-2 engines designed to work express passenger traffic on the Edinburgh to Aberdeen main line in September 1936. When Edward Thompson became the Chief Mechanical Engineer of the LNER in 1941, he very quickly decided the class required drastic modification. By 1944 they had been rebuilt with a Pacific wheel arrangement. No. 60506 spent the rest of its career based at New England MPD working on the East Coast Main Line. The class was an early casualty in the dieselisation programme, with No. 60506 being one of the last active members. It was withdrawn from service in April 1961.

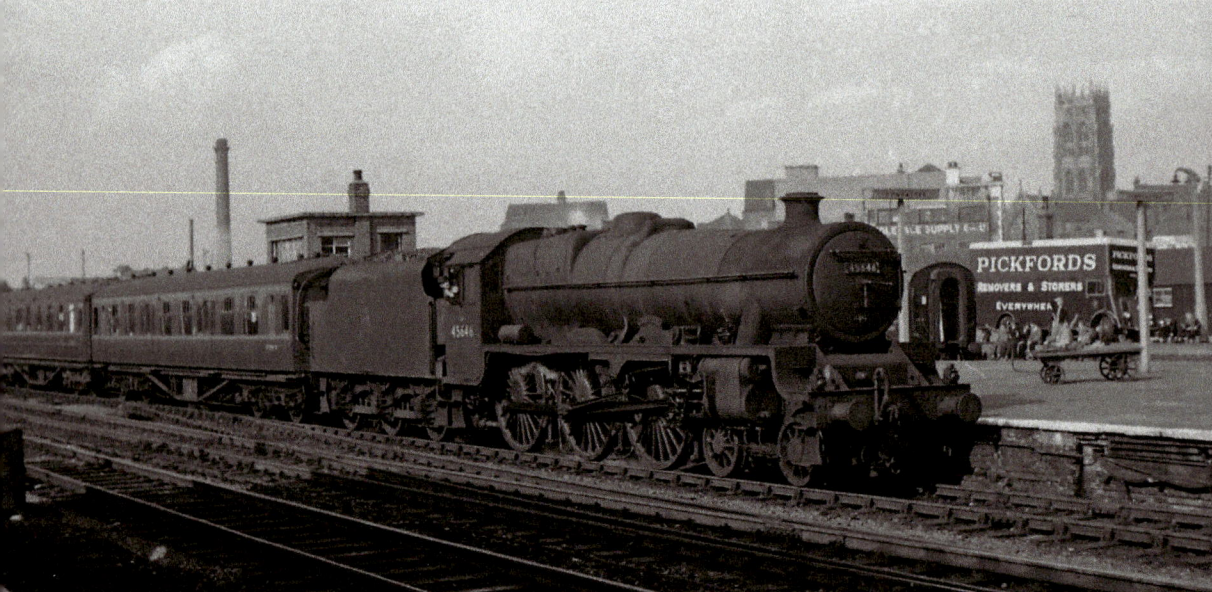

1 July 1961. LMS Jubilee 4-6-0 No. 45646 draws into Doncaster station.

LMS Jubilee 4-6-0 No. 45646 *Napier* was a resident of Farnley Junction MPD in Leeds and an unusual sight in Doncaster at the time. It was most probably a late replacement for a failed loco. A much more familiar sight in the background was one of the Pickford's Guy Otter pantechnicons, which were regularly parked on the cattle dock area and the throng of trainspotters who congregated there.

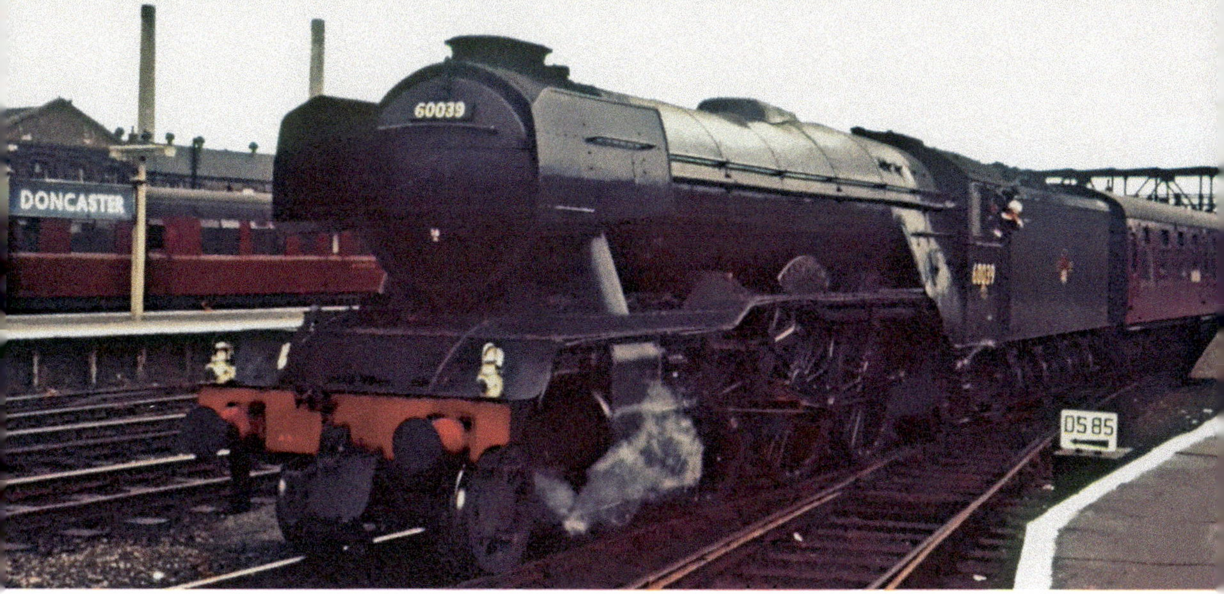

9 July 1961. LNER A3 4-6-2 No. 60039 departs from Doncaster station.

LNER A3 4-6-2 No. 60039 *Sandwich* gets its Kings Cross-bound train away from platform 4 at Doncaster station. It was a Kings Cross-allocated locomotive for the last six years of its working life, which was to last another eighteen months. It had worked the previous evening's Yorkshire Pullman to Leeds and had the name board reversed on the front buffer beam. The train is just passing under the footbridge, which connected the Plant works to the left of the station with the town. The bridge became a hive of activity at the end of the working day when many hundreds of workers walked and pushed their bicycles over it.

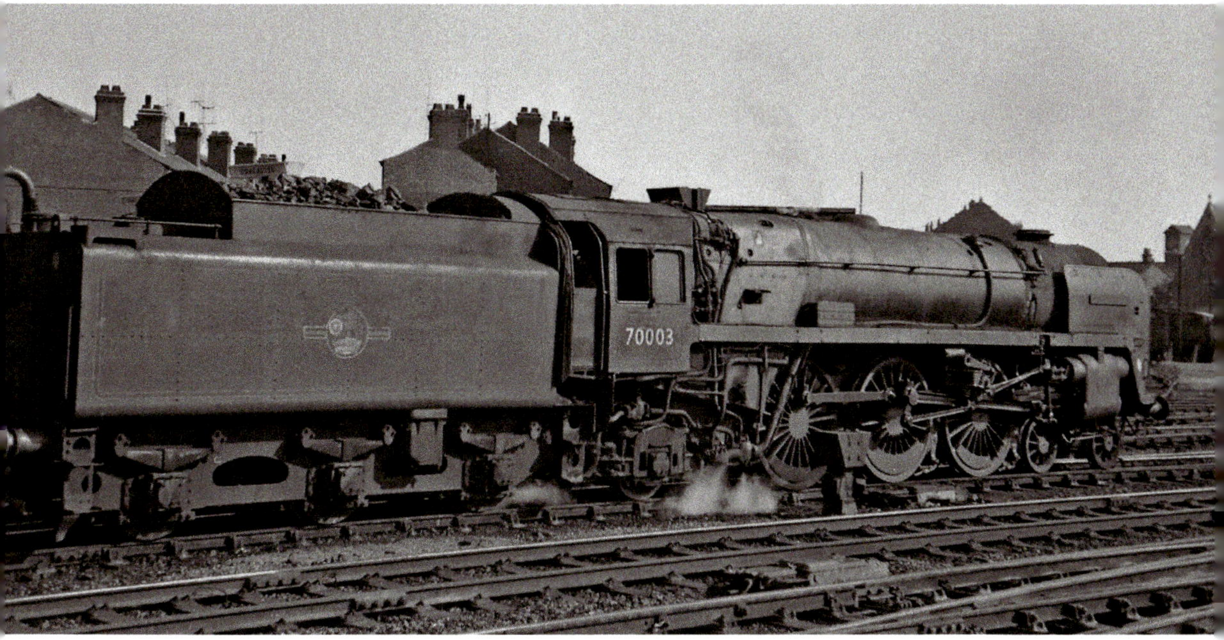

17 August 1961. BR Britannia 4-6-2 No. 70003 departs from Doncaster.

BR Britannia 4-6-2 No. 70003 *John Bunyan* departs southbound from Doncaster. March-allocated Britannias were a regular feature at Doncaster because they worked trains from East Anglia to York and they were also overhauled at the Plant. This train had originated from Newcastle with No. 70003 taking over haulage duties at York.

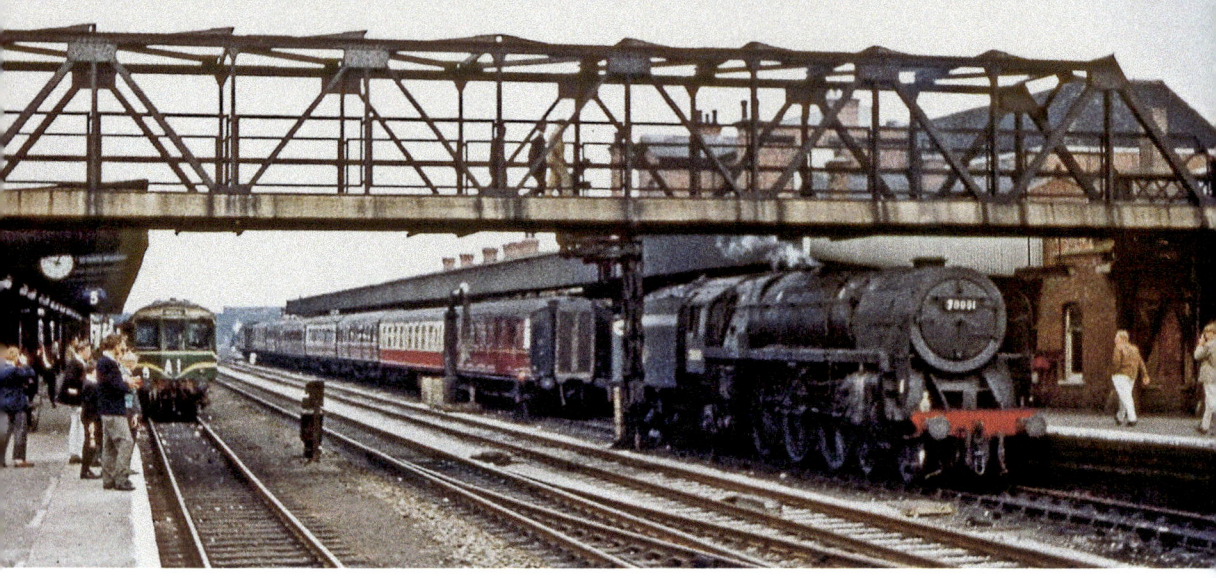

17 September 1961. BR Britannia 4-6-2 No. 70001 *Lord Hurcomb* attaches at Doncaster.

BR Britannia 4-6-2 No. 70001 *Lord Hurcomb* backs on to its train at Doncaster. The March-allocated locomotive would be taking the varied collection of stock back to East Anglia. The second coach, a BR Mark 1 in crimson and cream, was a rare sight in 1961 because repainting from this livery to maroon had begun five years earlier in 1956. The Birmingham Railway Carriage & Wagon Company four-car diesel multiple unit, later designated Class 104, on platform 5, was a sign of the changing face of the railway in the 1960s. The youthful group of trainspotters on this Sunday afternoon was very much smaller than the throng of young and old who gathered on Saturdays and who sometimes overwhelmed the platform capacity.

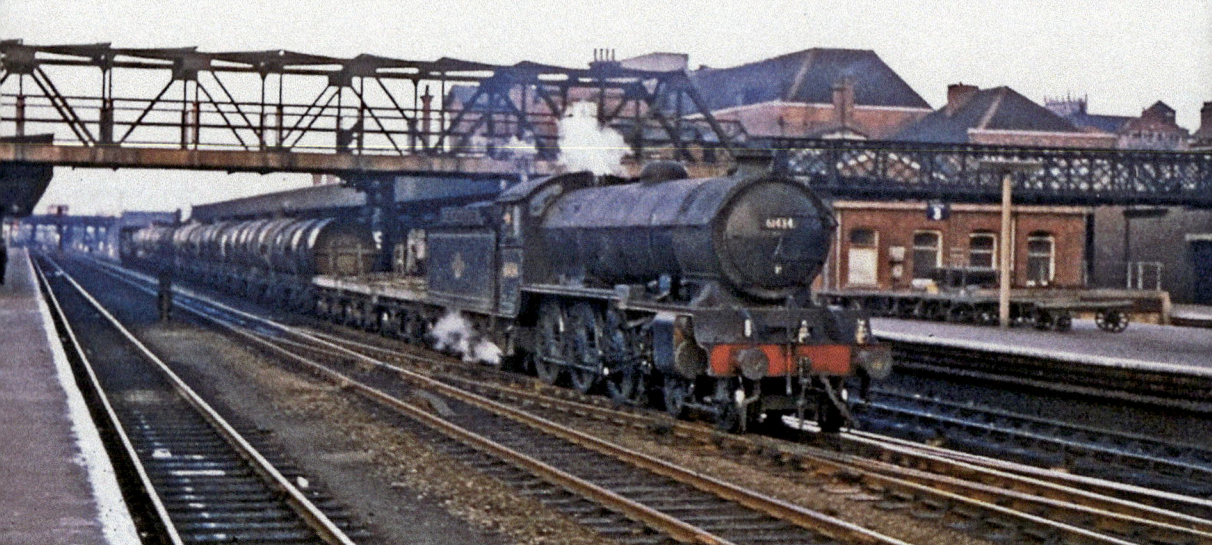

April 1962. LNER B16/3 4-6-0 No. 61434 rubbles through Doncaster.

LNER B16/3 4-6-0 No. 61434 heads a train of methanol from ICI on Teeside and takes the centre road south through the station. The use of three barrier wagons between the locomotive and the tankers give an indication of the care that was taken when transporting such hazardous materials. No. 61434 had been based at York for twenty-five years at the time of the photograph and its recent general overhaul would see it through to withdrawal in June 1964.

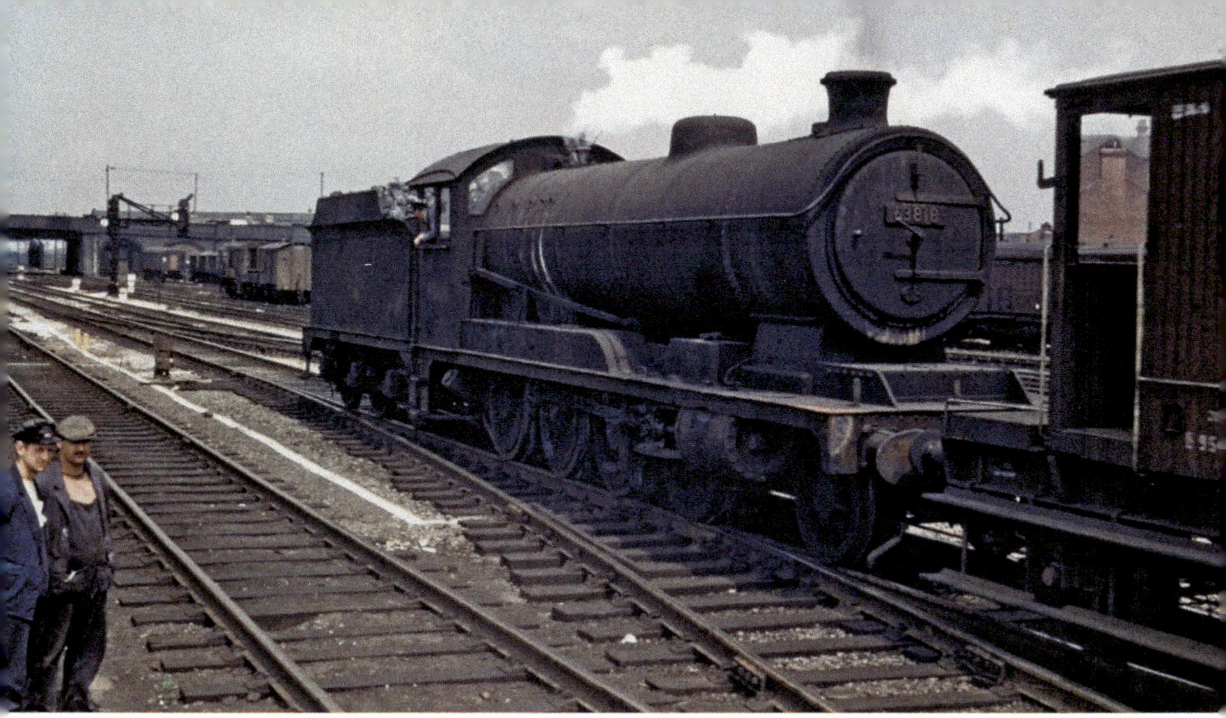

June 1962. LNER O4/8 2-8-0 No. 63818 trundles slowly through Doncaster station.

LNER O4/8 2-8-0 No. 63818 cautiously heads through Doncaster station with another string of coal empties for the South Yorkshire collieries north of Doncaster. Two enginemen watch its slow progress as they stand on the edge of one of the running lines.

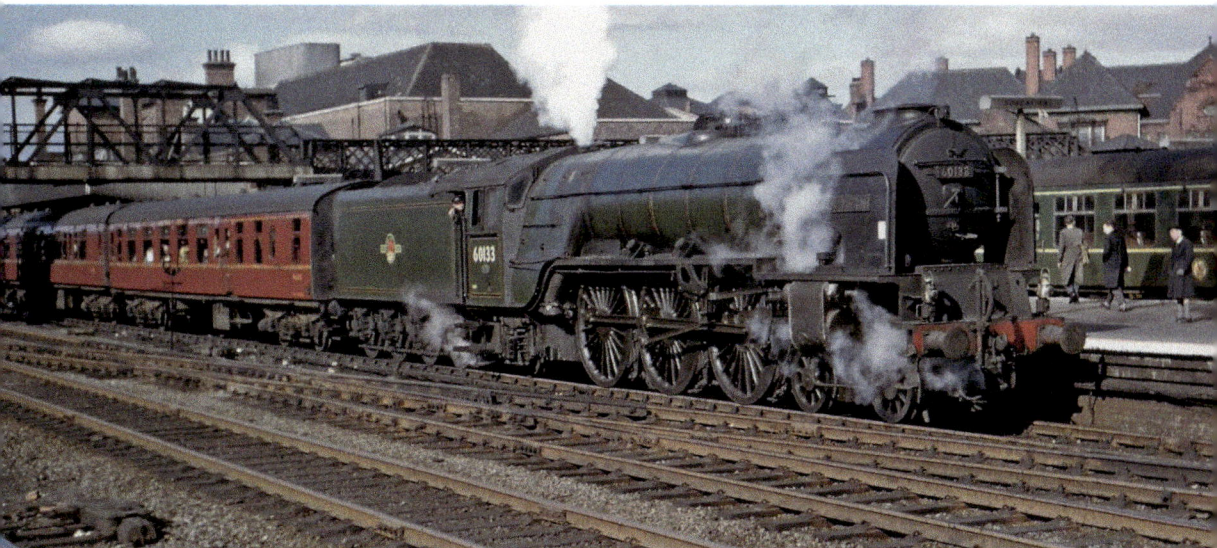

June 1962. LNER 4-6-2 A1 No. 60133 at Doncaster.

LNER 4-6-2 A1 No. 60133 *Pommern* entered service in October 1948 and for the first eighteen months of its career was allocated to Grantham MPD. During this short period it wore two different liveries; first LNER apple green and then BR lined blue livery. In June 1950 it was reallocated to Copley Hill MPD and remained loyal to that depot until its final move to Ardsley MPD from where it was withdrawn nine months later in June 1965. It spent most of its working life hauling expresses between Leeds and Kings Cross. It was regularly rostered for the Yorkshire Pullman but by 1965 this work was in the hands of the Deltics.

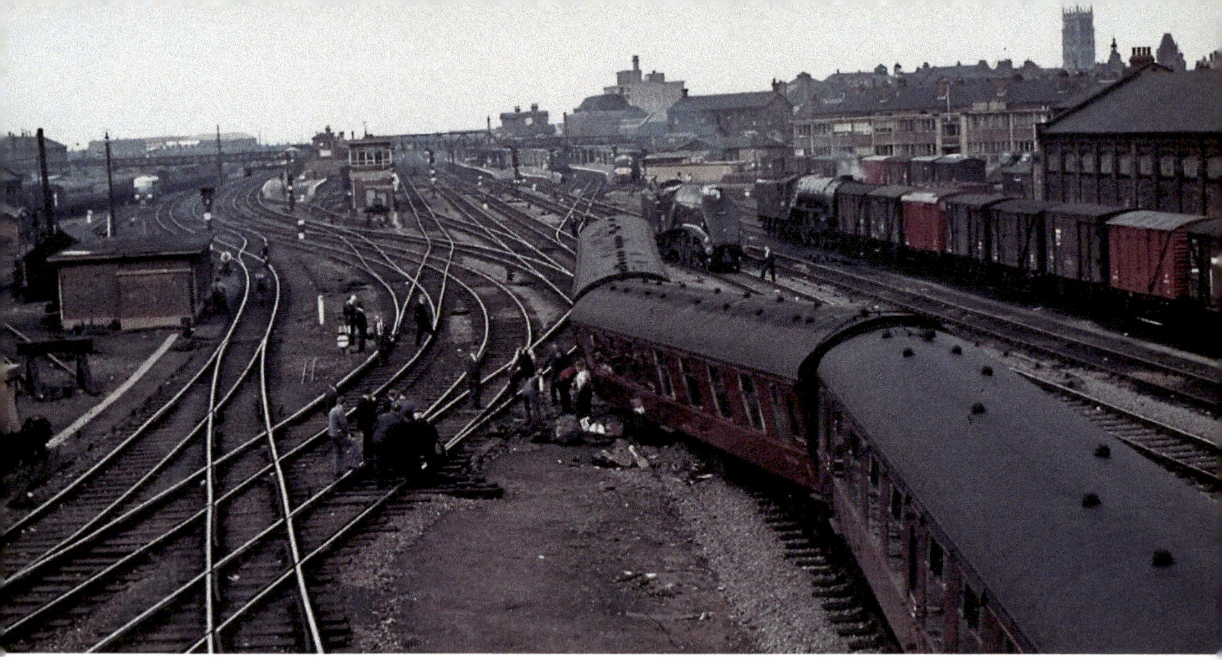

July 1962. LNER A4 4-6-2 No. 60022 passing derailment in Doncaster.

In July 1962, a derailment occurred with empty passenger stock being propelled into Doncaster station. It appears that the trailing bogie of the first coach and the lead bogie of the second coach had been derailed some distance before the train had come to a stop, as can be witnessed by the marks in the ballast alongside the train. LNER A4 4-6-2 No. 60022 *Mallard* reverses past the scene, wrong line working, and also past the A1 southbound standby locomotive as the emergency crews set about the task of re-railing. (L. Flint)

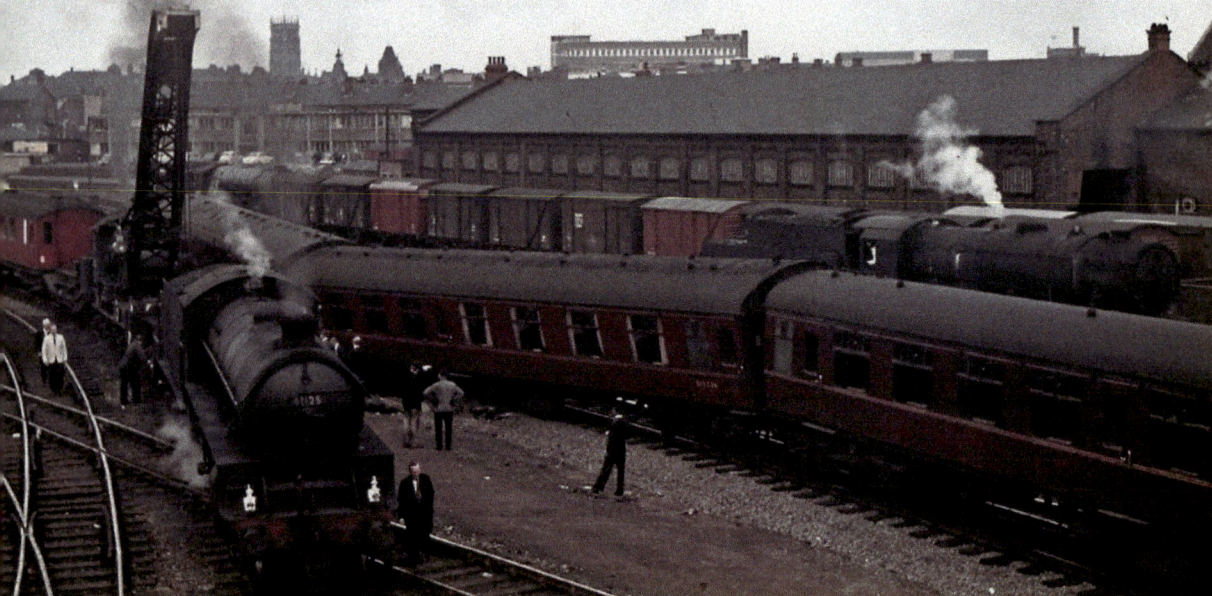

July 1962. LNER B1 4-6-0 No. 61125 attending the derailment in Doncaster.

LNER B1 4-6-0 No. 61125 arrived on the scene with the Doncaster breakdown train behind the derailment and a WD 2-8-0 passes working wrong line. The number of personnel on the scene and the complete absence of high-visibility clothing are remarkable by today's standards. The fact that traffic continued to pass the incident during re-railing operation also shows how procedures have changed. (L. Flint)

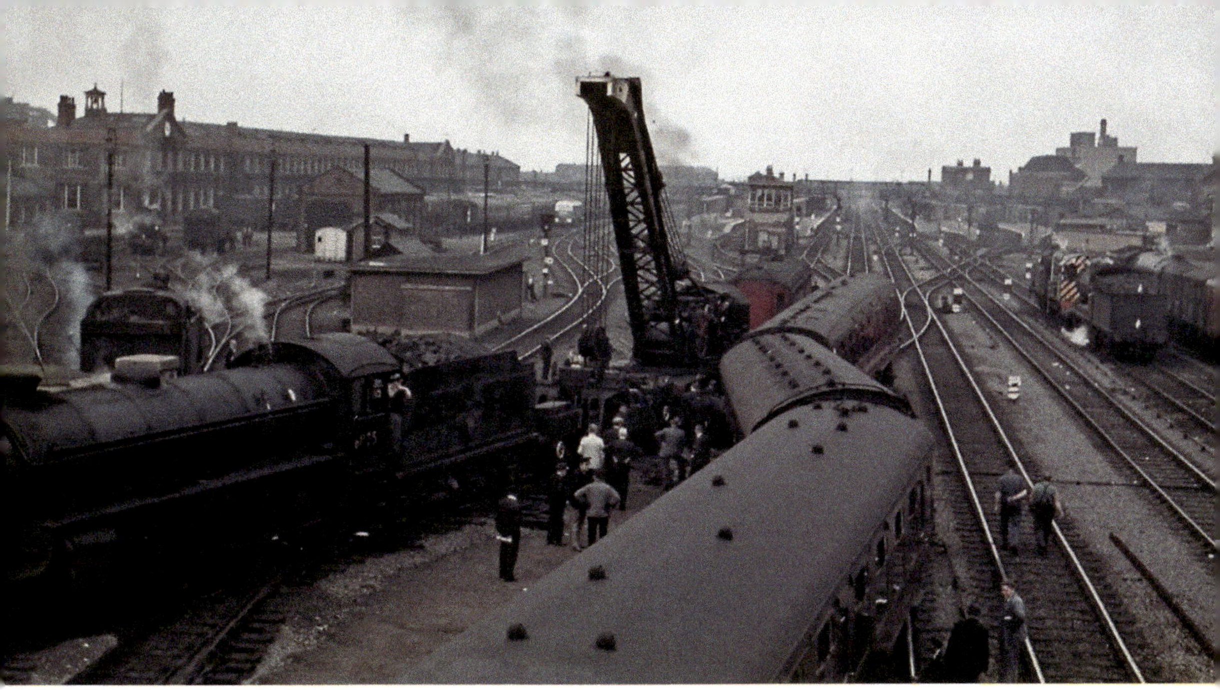

July 1962. LNER B1 4-6-0 No. 61125 at the derailment in Doncaster.

The disruption created by the derailment can be clearly seen with both down slow and fast mainlines blocked as well as both up and down lines to Sheffield. To the right of the derailment an O4 2-8-0 heads back to the Motive Power Depot with two repaired locomotives from the Plant. The southbound standby A1 4-6-2 Pacific waits behind the passing cavalcade. (L. Flint)

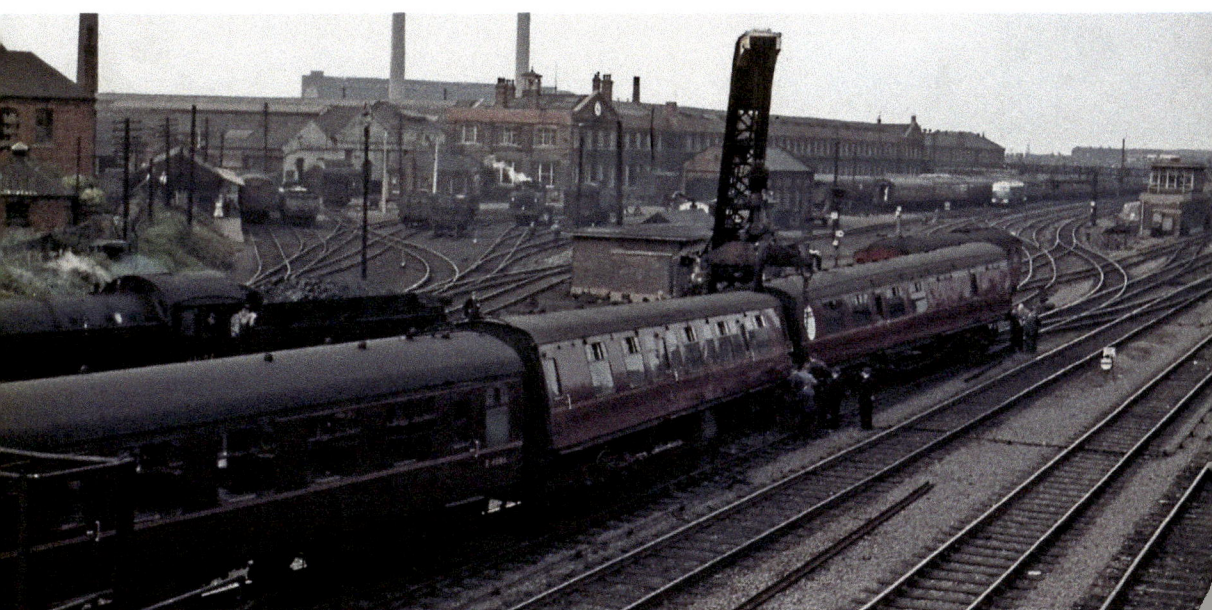

July 1962. Doncaster's steam crane attending the derailment.

The breakdown crane begins the lift of the leading coach to return it to the rails. In the background is the vast Plant Works with a number of J50 0-6-0Ts works shunters visible. In the distance one of the Plant's latest products, a Bo-Bo AL5 electric locomotive, is one of forty of the class built at Doncaster for the newly electrified West Coast Main Line. (L. Flint)

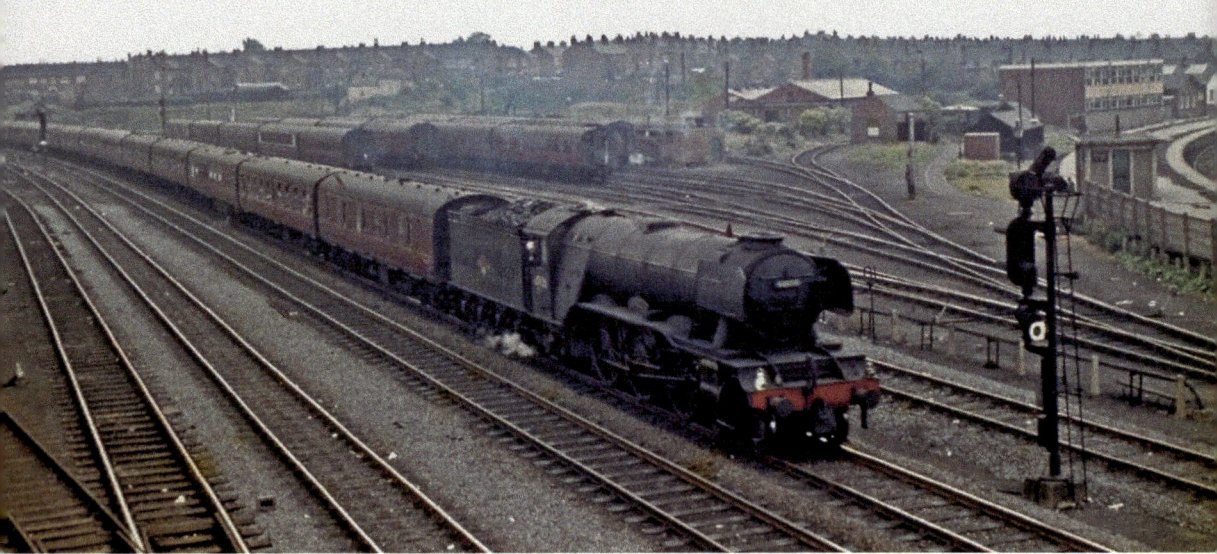

29 July 1962. LNER A3 4-6-2 No. 60106 approaching Doncaster.

LNER A3 4-6-2 No. 60106 *Flying Fox* had been built in April 1923 for the Great Northern Railway at Doncaster Plant as an A1. It was rebuilt as an A3 in April 1947 and was fitted with trough type smoke deflectors in October 1961. When it was finally withdrawn in December 1964, it was the LNER Pacific with the longest service life. It had worked mainly along the East Coast Main Line for forty-one years and eight months and covered well over 2.5 million miles, an amazing record and a great tribute to its designer Nigel Gresley.

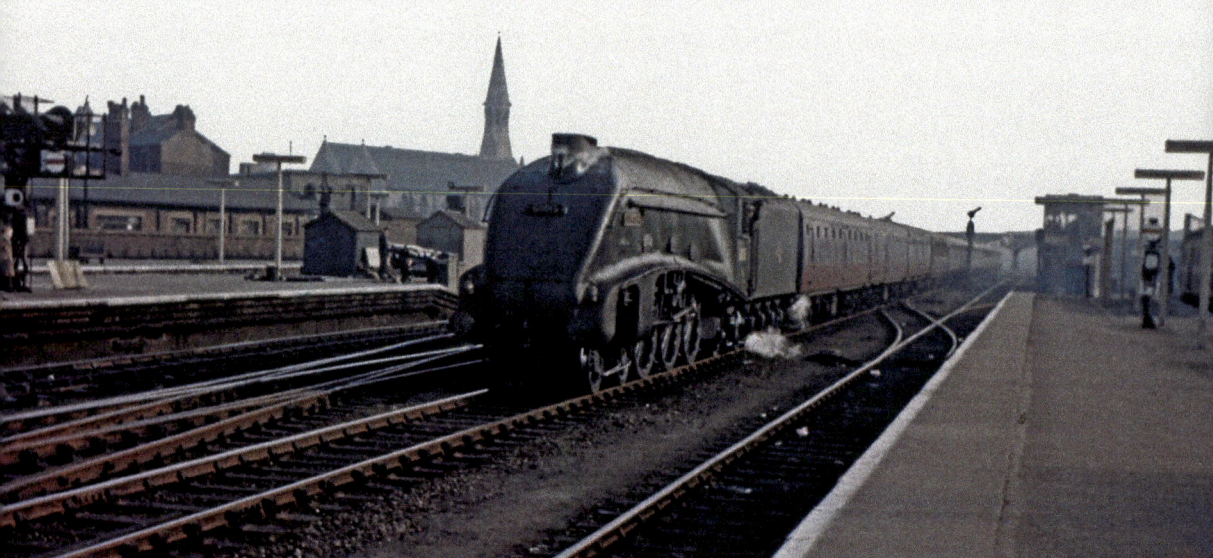

September 1962. LNER A1 4-6-2 No. 60015 rushes through Doncaster station.

LNER A4 4-6-2 No. 60015 *Quicksilver* blasts its chime whistle as it hurries a northbound express through Doncaster station. A small boy gazes on at the spectacle, no doubt in wonder at the locomotive's majestic progress. No. 60015 had been built at the Plant in Doncaster in September 1935 and had spent much of its twenty-seven-year career allocated to Kings Cross Top Shed. Its days of dominating the East Coast Main Line expresses would soon come to an end, as it was withdrawn in April 1963 and scrapped soon after.

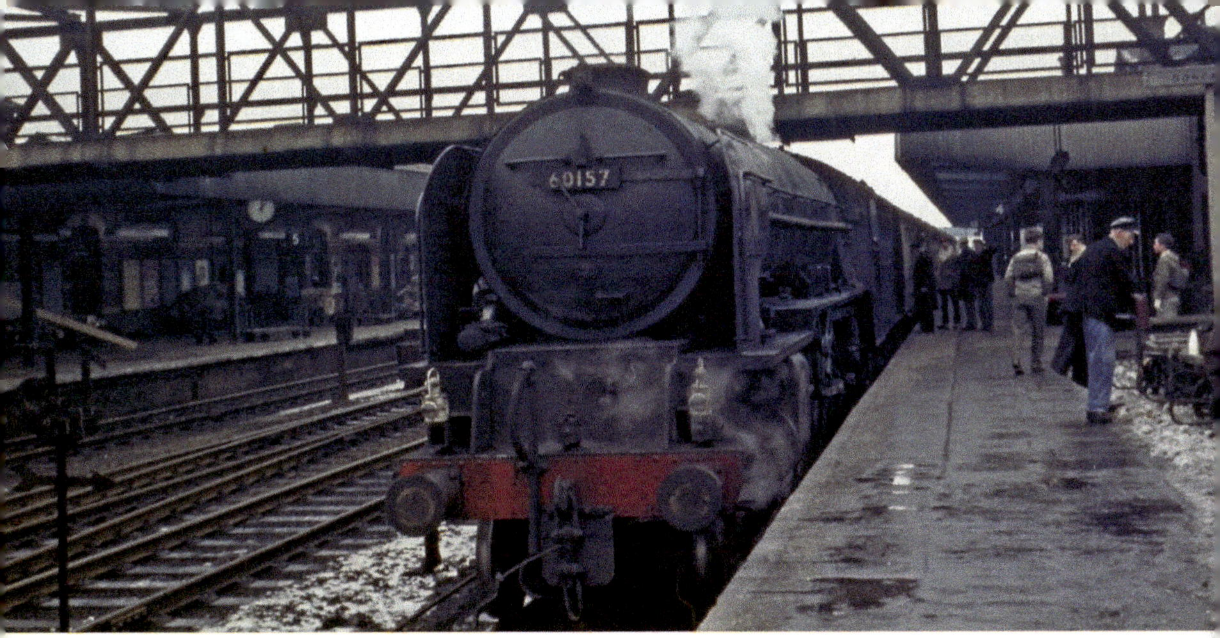

6 January 1963. LNER A1 4-6-2 No. 60157 couples onto its train at Doncaster.

LNER A1 4-6-2 No. 60157 *North Eastern* was a Doncaster-allocated locomotive that had two more years service before withdrawal. It is seen at Doncaster station being coupled to a London-bound train on a bitterly cold day. January 1963 was the coldest month of the twentieth century with temperatures dropping to -19° C and much of England and Wales was snow-covered. It was correctly known as 'The Big Freeze' because it was into March before outdoor temperatures reached above freezing point. The effects on everyday life were far reaching and on the railways the new diesel locos found the conditions very difficult. Steam locos appeared to cope better, and most trains seemed to be steam hauled. The snow around the station was not so deep but the cold was biting.

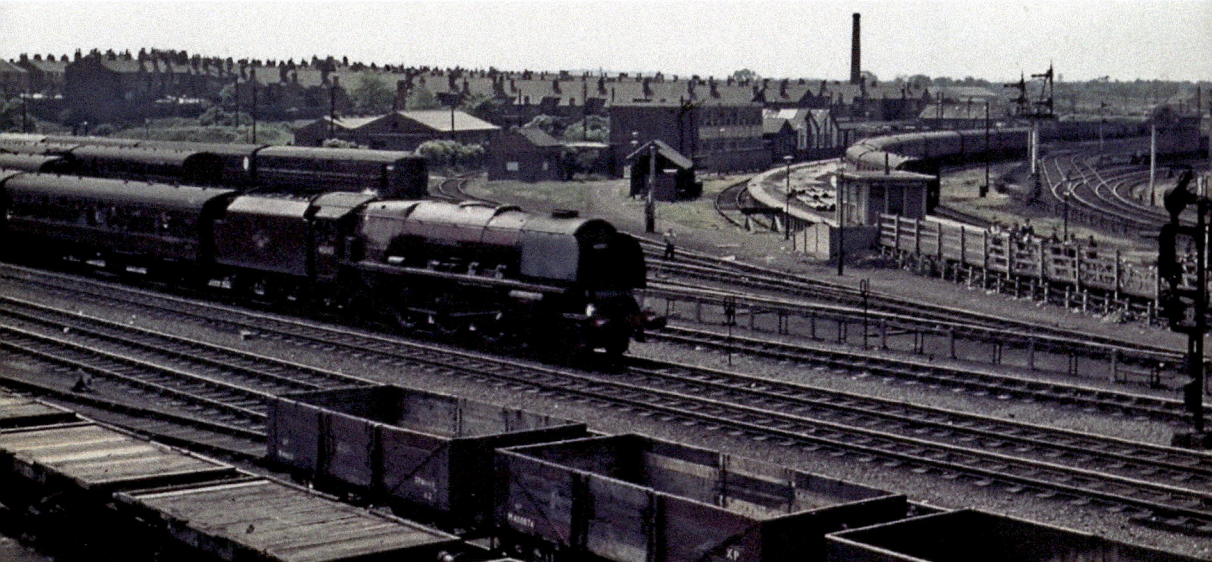

9 June 1963. LMS Duchess 4-6-2 No. 46245 arrives in Doncaster.

LMS Duchess 4-6-2 No. 46245 *City of London* at the head of the Home Counties Railway Club enthusiasts special from London Kings Cross to Doncaster. These special trains had become a regular feature of weekend workings in the 1960s. They were a magnet to local trainspotters as it was a chance to see some unusual exotic motive power and a chance to illicitly join a large group of visitors to look around the shed and the Plant. (L. Flint)

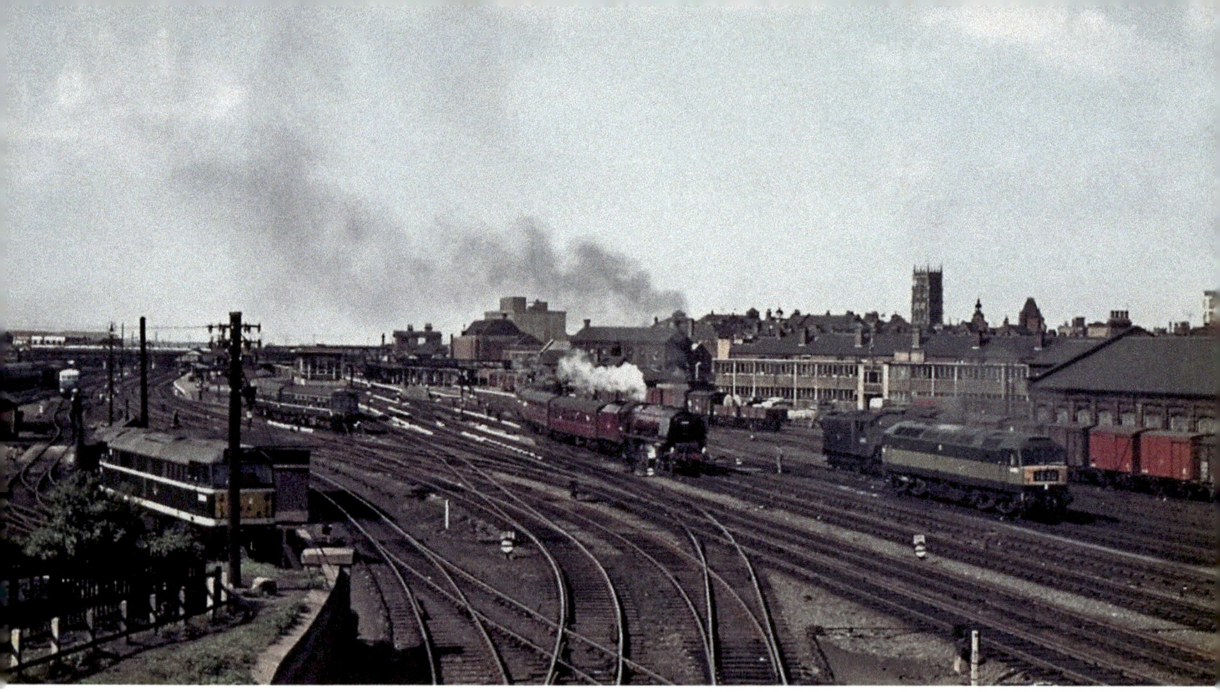

9 June 1963. LMS Duchess 4-6-2 No. 46245 departs south from Doncaster.

LMS Duchess 4-6-2 No. 46245 *City of London* confidently gets the twelve-coach Home Counties Railway Club enthusiasts special under way from Doncaster back to London Kings Cross. The new and changing face of the railway scene in the 1960s is very evident; Brush Type 2 and Type 4 diesels, a diesel multiple unit and a new electric locomotive dominate in this image. (L. Flint)

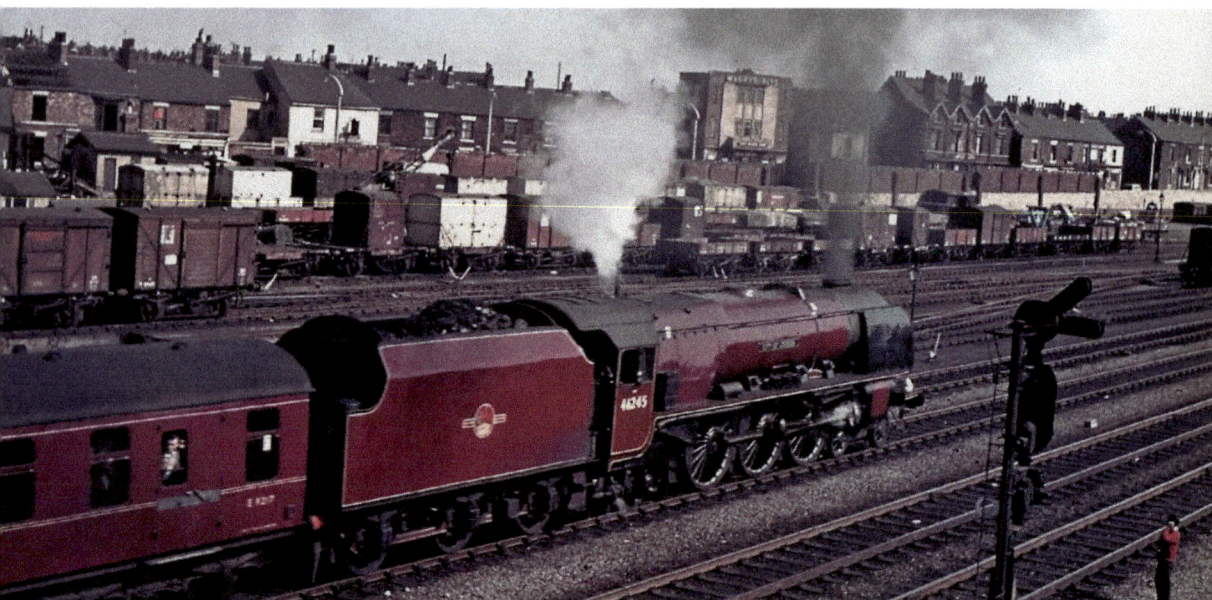

9 June 1963. LMS Duchess 4-6-2 No. 46245 leaves Doncaster for Kings Cross.

LMS Duchess 4-6-2 No. 46245 *City of London* heads the Home Counties Railway Club twelve-coach enthusiasts special back to London Kings Cross from Doncaster. No. 46245 had been a London-based locomotive for its entire twenty-one-year career but within eighteen months of this image the loco would be consigned to the scrapyard. (L. Flint)

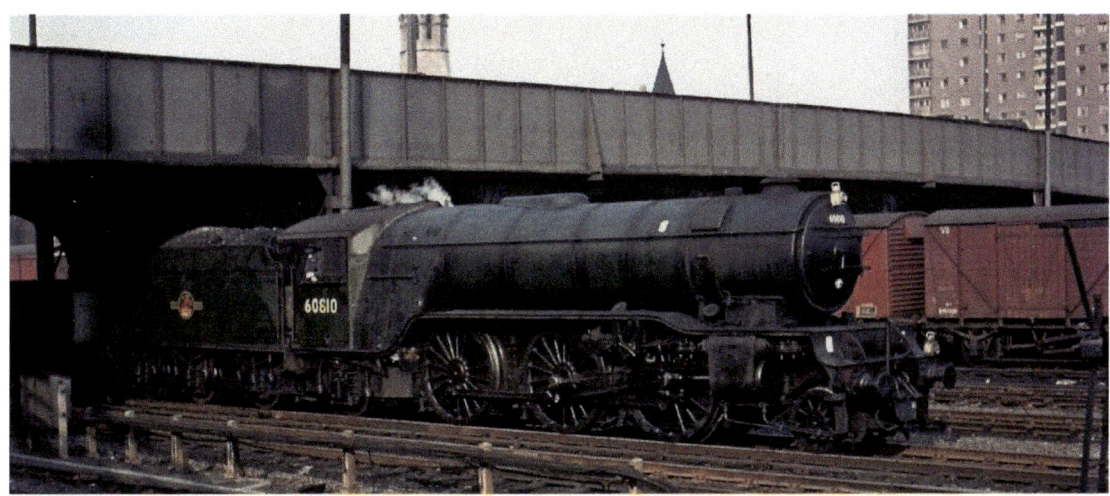

9 June 1963. LNER V2 2-6-2 No. 60810 heads under St James' Bridge in Doncaster.

LNER V2 2-6-2 No. 60810 heads under St James' Bridge with a southbound string of iron-ore empties, which were returning to Highdyke, 4 miles south of Grantham. At Highdyke there was a junction with the freight-only branch line to Stainby and Sproxton, which served the many ironstone quarries in the area where the empty wagons would receive their next load of ore. (L. Flint)

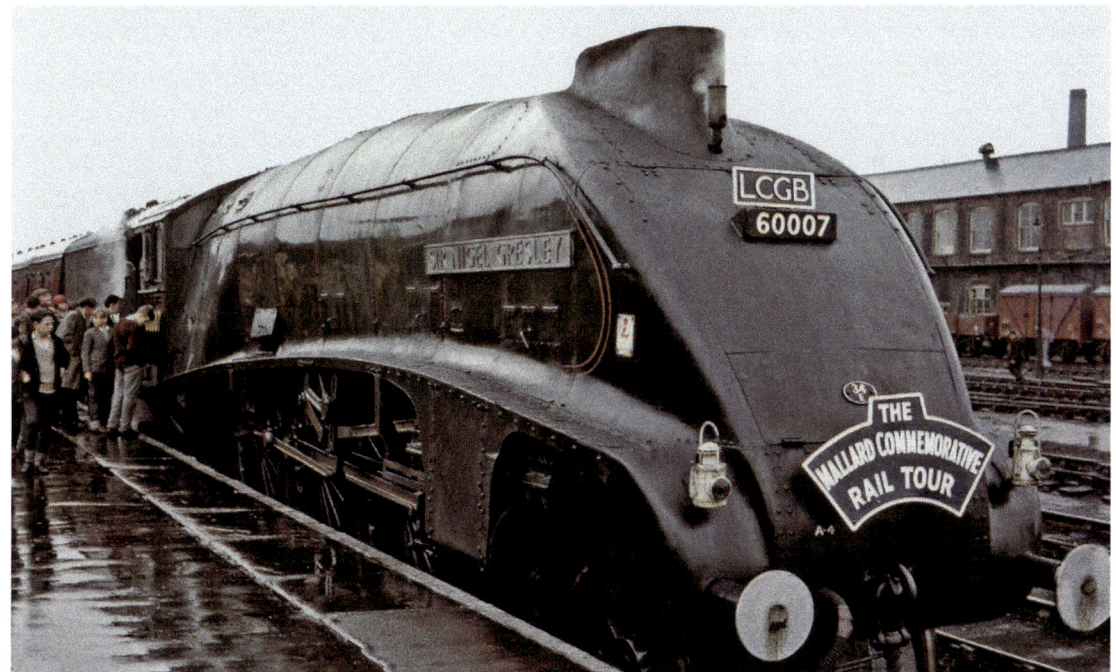

6 July 1963. LNER A4 4-6-2 No. 60007 attracts a crowd in Doncaster.

LNER A4 4-6-2 No. 60007 *Sir Nigel Gresley* at Doncaster with the LCGB Mallard Commemorative Rail Tour. The tour, which started at Kings Cross, was to commemorate the 25th anniversary of the world speed record achieved by *Mallard*. It had travelled to Doncaster via Cambridge, March and Gainsborough and visits were made to Doncaster Plant, York railway museum and depot. The return journey from York to Kings Cross was made down the East Coast Main Line and No. 60007 was recorded with a maximum speed of 102 mph descending Stoke bank.

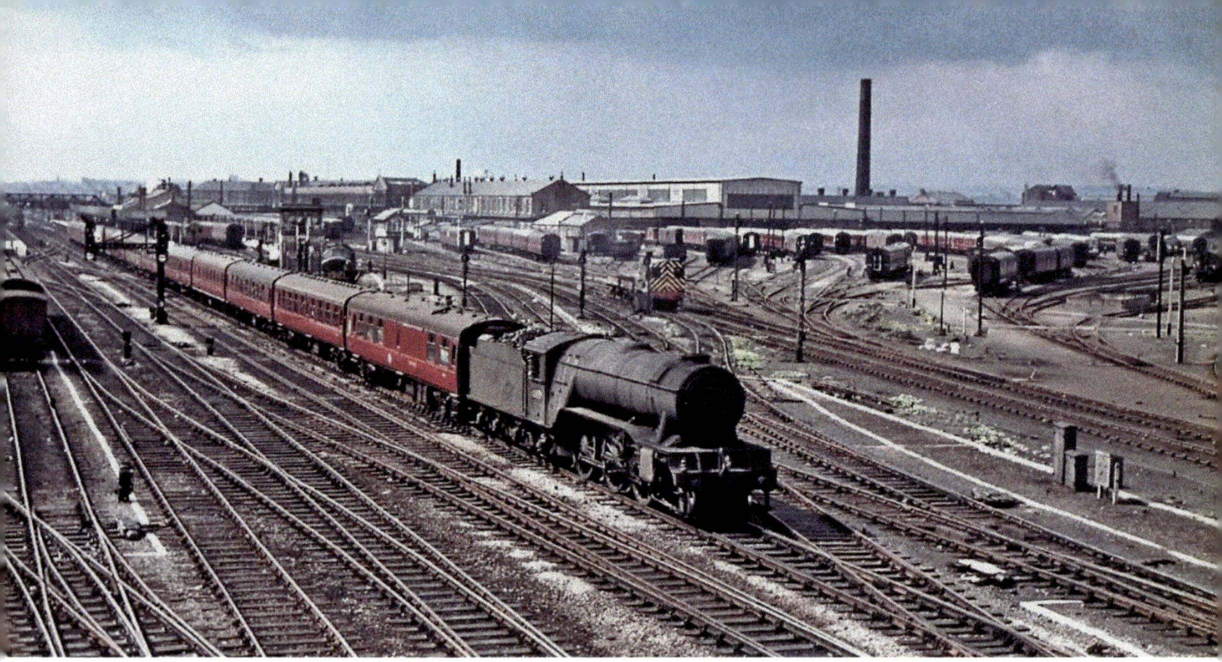

25 July 1963. LNER V2 2-6-2 No. 60899 rushes through Doncaster station.

LNER V2 2-6-2 No. 60899 rushes through Doncaster station from North Bridge with a northbound express. The centre roads through Doncaster station were reserved for non-stop express passenger services and between these the mundane, slow-moving freight trains. V2 2-6-2s shared the express work with the more glamorous LNER Pacifics, particularly the A4s. To the right of the train is the vast complex of sidings and buildings that formed part of the Plant works and, most prominent in this view, the carriage and wagon works. (L. Flint)

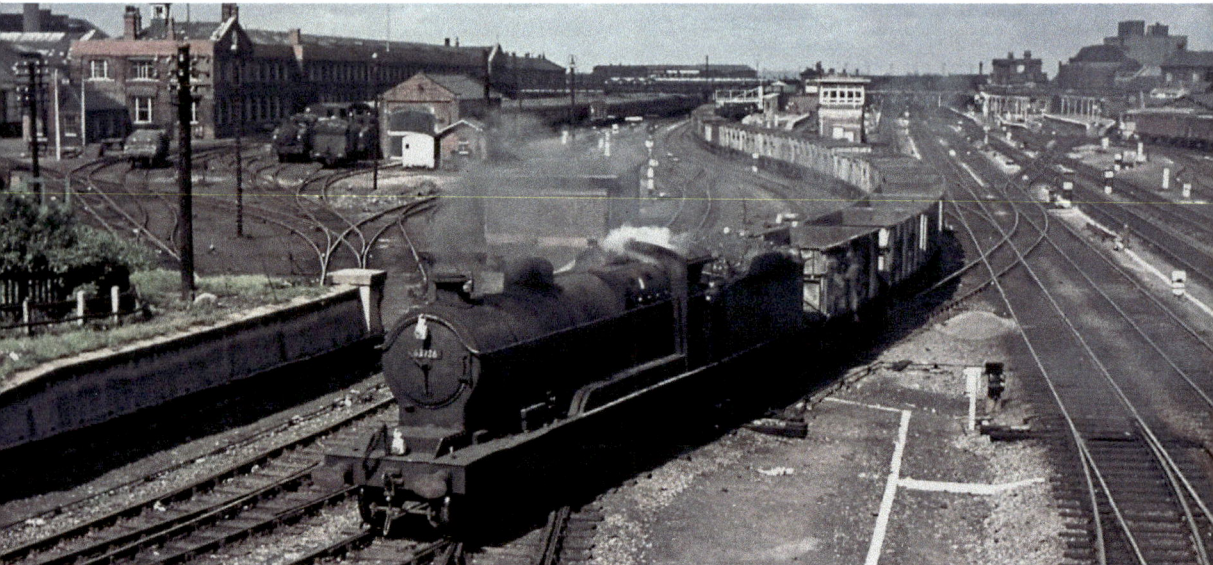

24 August 1963. LNER O4 2-8-0 No. 63736 heads through Doncaster.

LNER O4 2-8-0 No. 63736 trundles a long string of empty mineral wagons on the freight avoiding line around the back of Doncaster station. The train was heading over the Sheffield junction, no doubt to one of the many collieries along the route. No. 63736 was one of the original Great Central locomotives delivered in August 1912 and had been allocated to Retford Thrumpton depot for over twenty years. This working may well have been the last duty it performed as it was withdrawn five days later. (L. Flint)

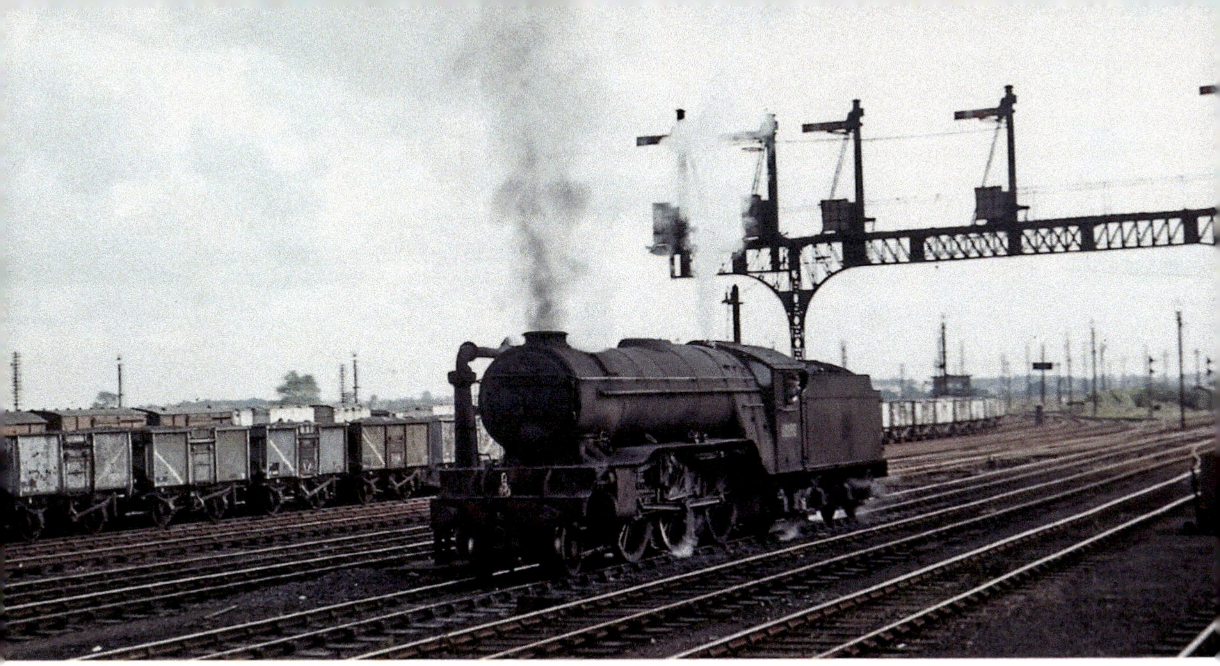

31 August 1963. LNER V2 2-6-2 No. 60880 pauses at signals in Doncaster.

LNER 4-6-2 V2 No. 60880 had been built at Doncaster Plant in September 1940 and had spent twenty of its twenty-three-year career allocated to Doncaster MPD. In the summer of 1961, it was one of the V2s to be equipped with a double chimney and Kylchap cowl, which improved its performance considerably but did little to delay its ultimate demise. It was in its last month of service as it waits at signals near Decoy No. 2 signal box during a light engine movement and was photographed from a passing down train.

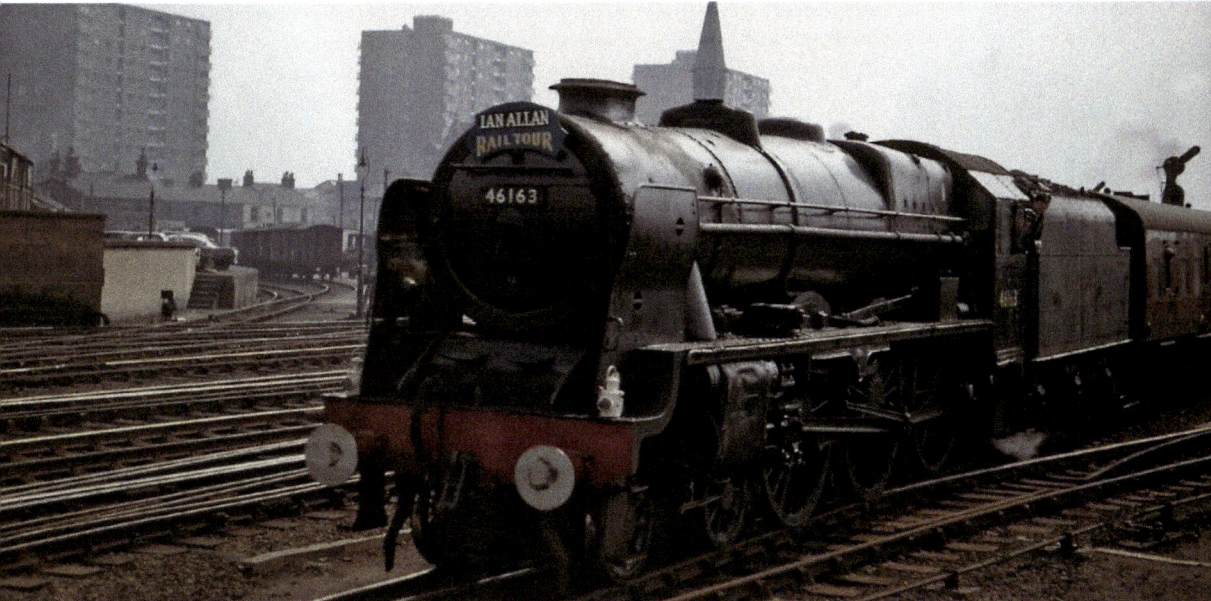

7 April 1964. LMS Royal Scot 4-6-0 No. 46163 heads a special into Doncaster.

LMS Royal Scot 4-6-0 No. 46163 brings its Ian Allan Locospotters' Club special into Doncaster. The train had originated from London Paddington and was hauled to Leicester Central up the Great Central line by GWR Castle 4-6-0 No. 7029 *Clun Castle*. There was an engine change at Leicester with No. 46163 *Civil Service Rifleman* taking over. The loco had been spruced up for this swansong duty but was withdrawn a few months later.

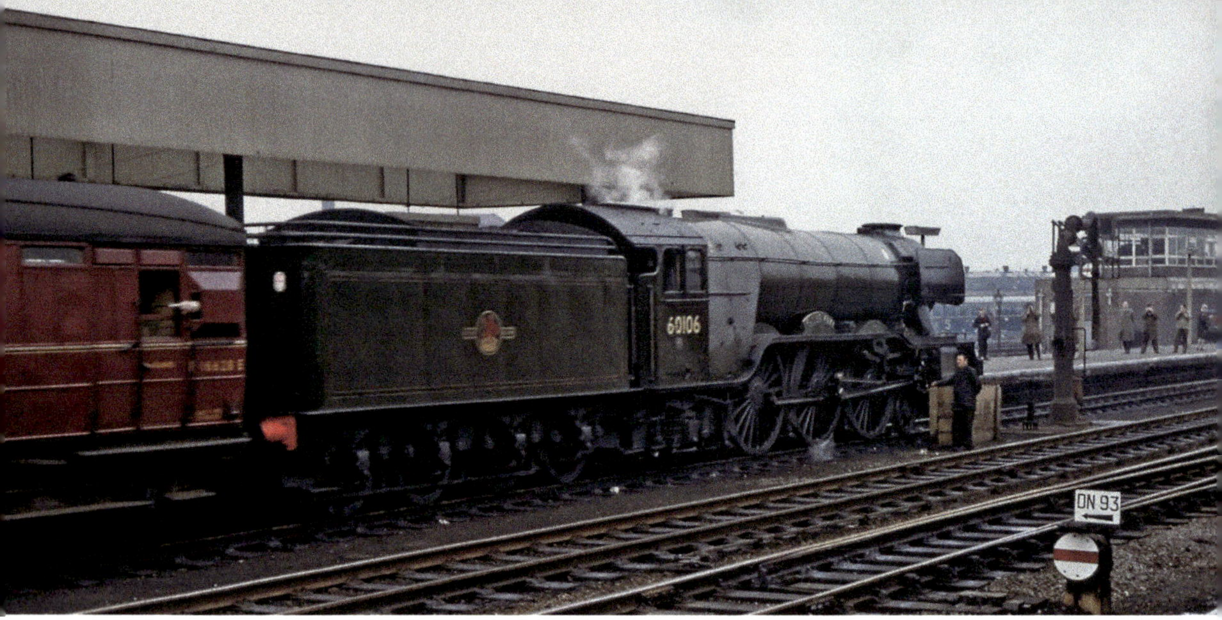

2 May 1964. LNER A3 4-6-2 No. 60106 at Doncaster station with a special.

LNER A3 4-6-2 No. 60106 *Flying Fox* arrives at Doncaster station with the Gresley Society special, the *North Eastern Flyer*. It hauled the special from Kings Cross to Doncaster and return with No. 4472 *Flying Scotsman* heading the train to Darlington and returning it to Doncaster. No. 60106 was recorded at 98 miles per hour on the descent of Stoke Bank on the return journey. It was withdrawn in December 1964 after having maintained schedule on many substitute appearances for Deltic diesels during its final year of service.

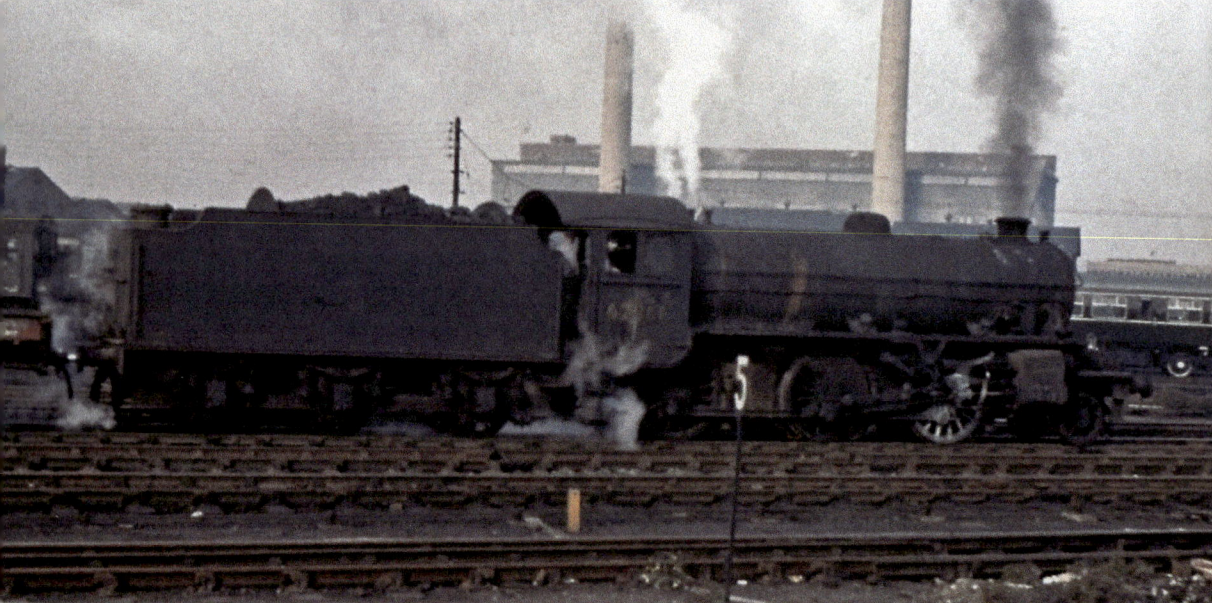

May 1964. LNER K1 2-6-0 No. 62066 restarts its train through Doncaster.

LNER-designed K1 2-6-0 No. 62066 had been built by the North British Locomotive Company in Glasgow in January 1950 for the relatively new British Railways. It had been allocated to Doncaster MPD in March 1963 and was to be withdrawn from there in January 1965. It was seen in typical careworn condition working a goods train northbound through Doncaster station. In the background are the two huge chimneys and massive bulk of Doncaster power station. This structure has long since disappeared and a prison has been built on the site.

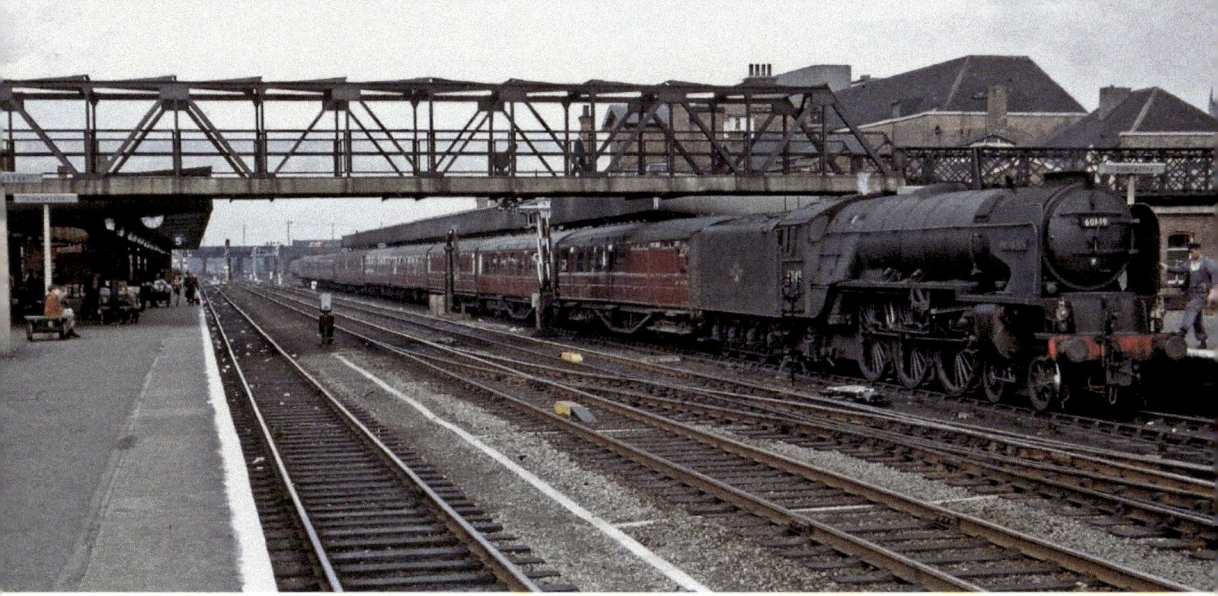

September 1964. LNER A1 4-6-2 No. 60149 at Doncaster with a Leeds to Kings Cross.

LNER designed 4-6-2 A1 No. 60149 *Amadis* had been built at Darlington and entered service in May 1949. In common with a lot of other A1s, in its first few years it had worn three liveries – apple green, BR blue and Brunswick green. It had worked from Kings Cross and Grantham depots before reallocation to Doncaster in September 1958. It was seen in Doncaster station working a Leeds to Kings Cross train just days before withdrawal.

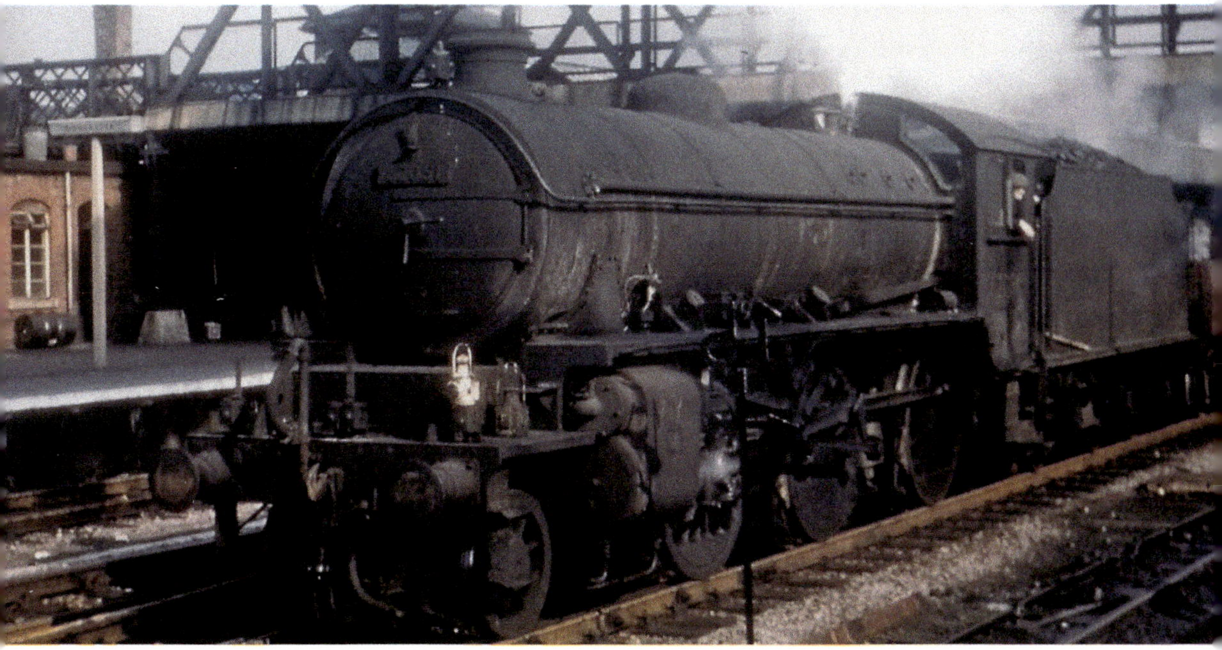

September 1964. LNER K1 2-6-0 No. 62051 heads south through Doncaster.

LNER-designed K1 2-6-0 No. 62051 in very work-stained condition heads a mixed goods train through Doncaster. It had been a Doncaster engine since late in 1962 and would be withdrawn in January 1965. The train is on the southbound centre road of the station, which was more regularly associated with non-stop expresses storming through very quickly. Slow, plodding goods trains used the lines also and were often held at signals within the station confines.

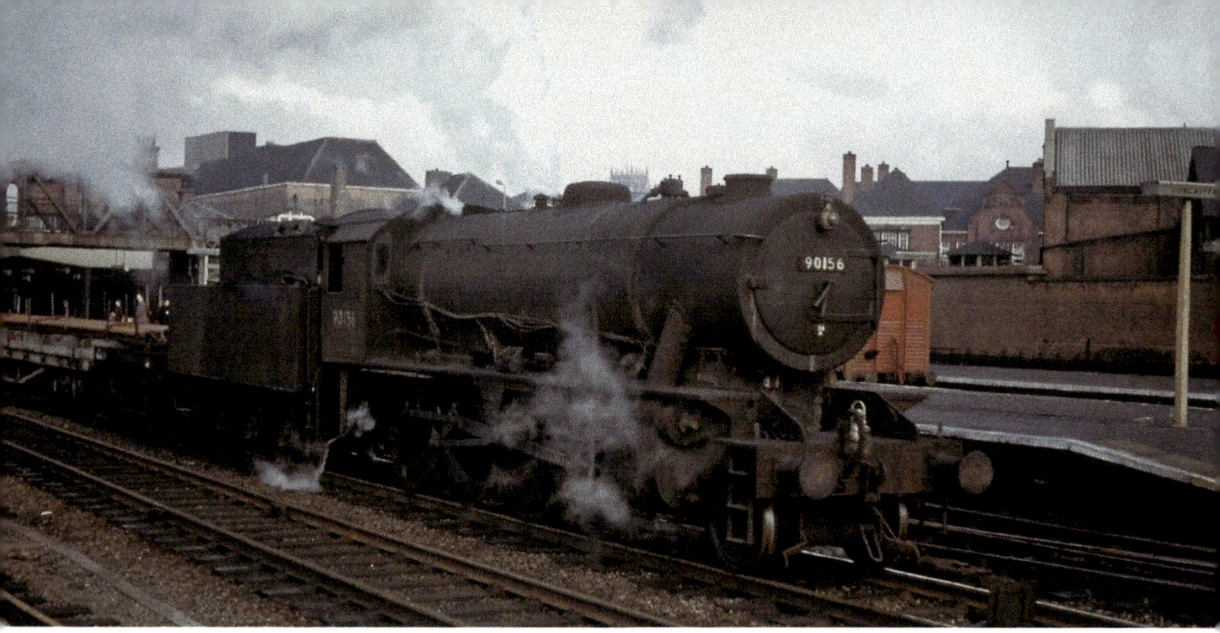

October 1964. WD 2-8-0 No. 90156 rattles through Doncaster.

WD 2-8-0 No. 90156 had been built by the North British Locomotive Company of Glasgow in June 1943 for the War Department. It was immediately loaned to the LNER until called for service in France in January 1945. After its return from France it was overhauled, modified and loaned once again to the LNER. The loco was acquired by British Railways at the time of nationalisation and by 1961 had become part of the stud of WDs allocated to Doncaster MPD. No. 90160 remained at Doncaster until the depot closed to steam in April 1966 and it was withdrawn and scrapped at Draper's of Hull.

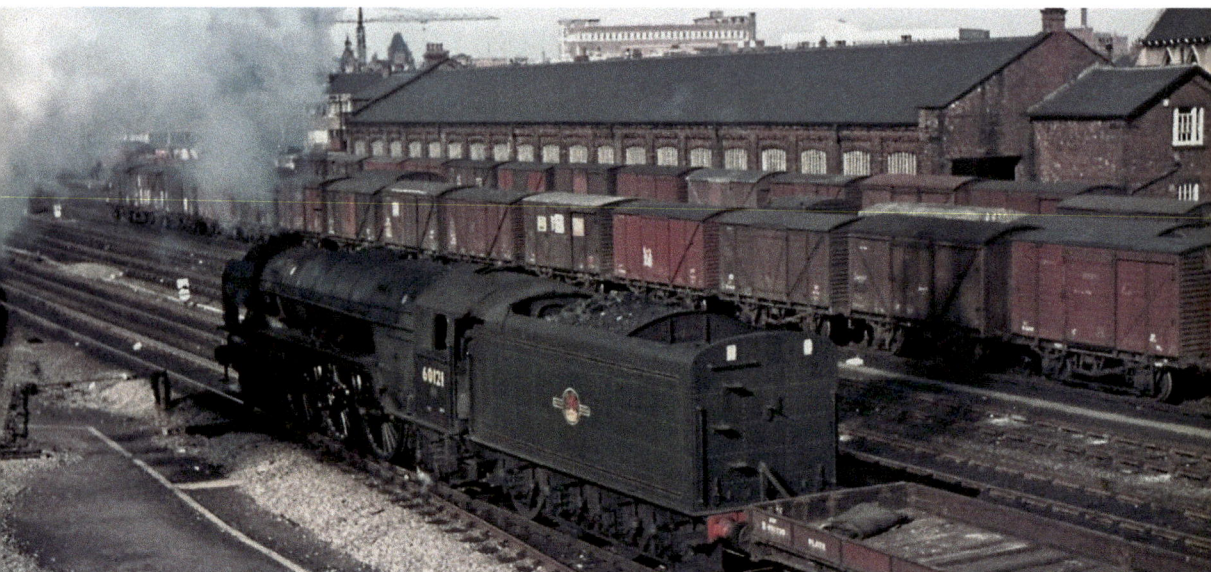

April 1965. LNER A1 4-6-2 No. 60121 heads north under St James' Bridge.

LNER A1 4-6-2 No. 60121 *Silurian* was a York MPD-based locomotive throughout its career. It had been built at Doncaster and entered service in late December 1948. During its sixteen-year working life it would have raced up and down the East Coast Main Line with express passenger trains, but by 1965 it had been relegated to hauling humble goods trains. Within six months this fine-looking loco would be sent to T. W. Ward of Killamarsh to feed the Sheffield blast furnaces. (L. Flint)

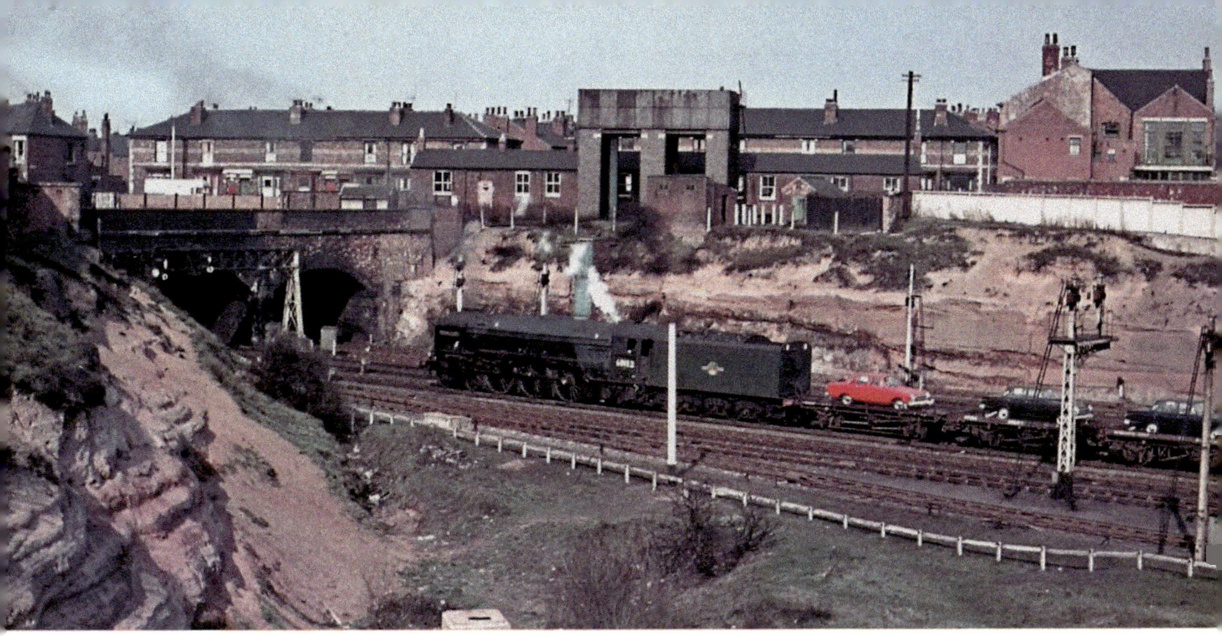

April 1965. LNER A1 4-6-2 No. 60152 approaches Balby Bridge, Doncaster.

LNER A1 4-6-2 No. 60152 *Holyrood* had spent most of its sixteen years of service as a Scottish-based locomotive and made only rare visits to Doncaster for general overhauls at the Plant. The last of these was in May 1963 and a year later it would be transferred to York MPD for the last nine months before withdrawal. It was seen heading north with a car transporter train loaded with brand new Ford Cortina Mark 1s from Dagenham. Ironically, the Ford plant in Doncaster, which produced over 100,000 vehicles, was being closed at this time because the government had given Ford financial incentives to open the Halewood plant on Merseyside rather than expand Doncaster. (L. Flint)

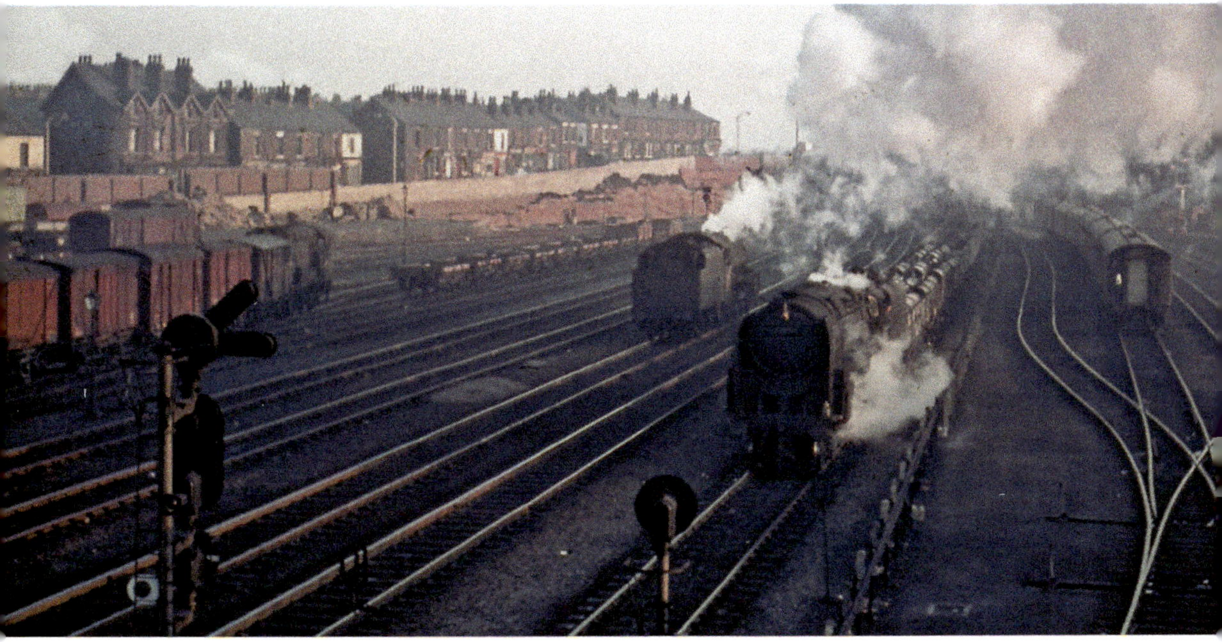

April 1965. BR 9F 2-10-0 No. 92146 heads a goods train under St James' Bridge.

BR 9F 2-10-0 No. 92146 accelerates a northbound fitted goods train towards St James' Bridge as it passes LMS Jubilee 4-6-0 No. 45593 *Kolhapur*. It had arrived on a Leeds to Doncaster local and having left its stock in the Garden Sidings to the right was making its way to the motive power depot for servicing and turning.

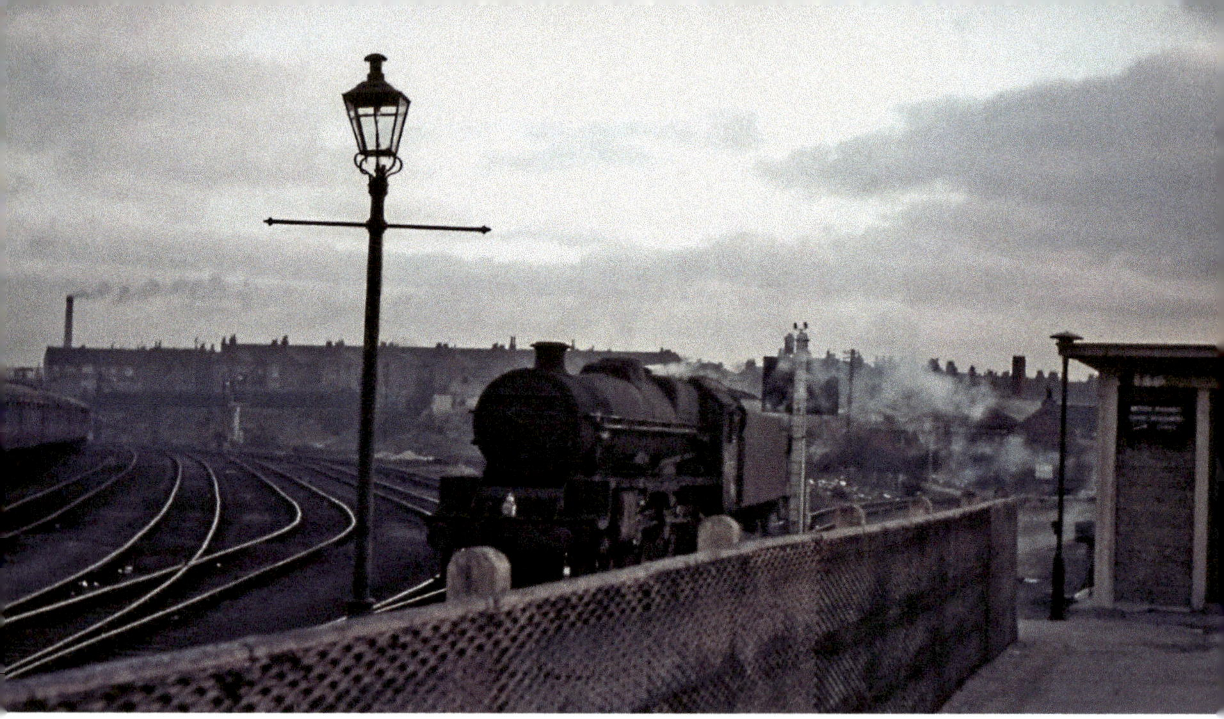

April 1965. LMS Jubilee 4-6-0 No. 45593 eases past Garden Sidings.

LMS Jubilee 4-6-0 No. 45593 *Kolhapur* having been serviced and turned runs into Garden Sidings to collect its stock for the return working to Leeds. The photographer is standing on the platform of the long-closed Doncaster excursion station, which had been built specifically to cope with passenger traffic to Doncaster racecourse. (L. Flint)

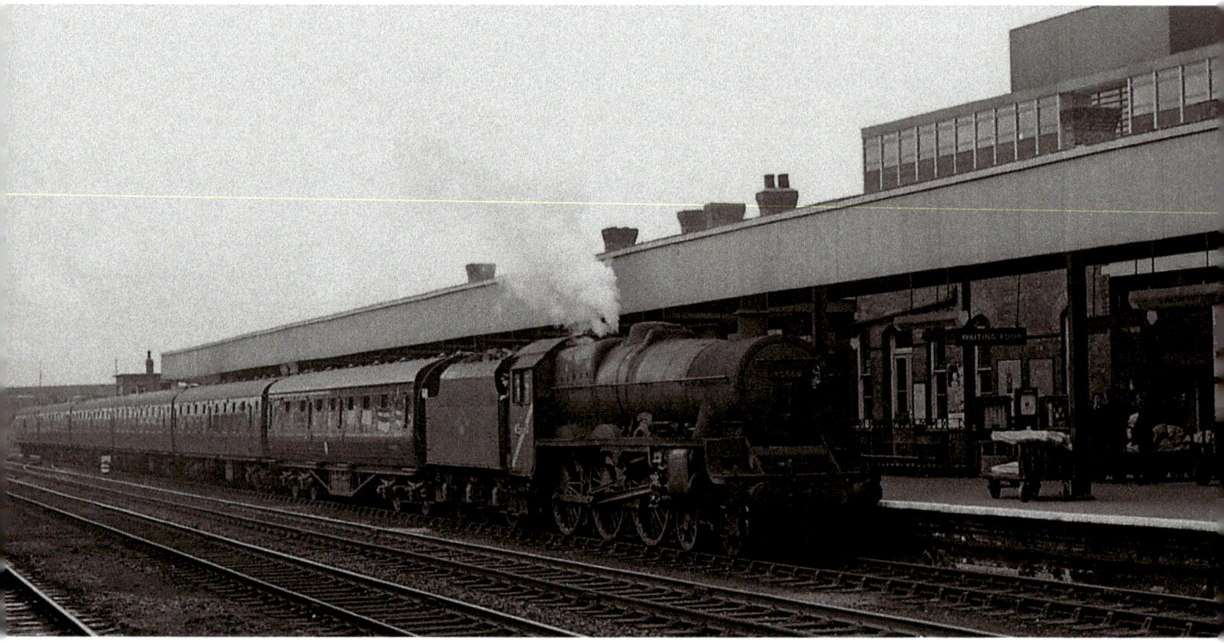

July 1965. LMS Jubilee 4-6-0 No. 45660 arrives at Doncaster.

LMS Jubilee 4-6-0 No. 45660 *Rooke* draws into Doncaster station with a Leeds to Doncaster local. Jubilees had become the regular motive power on the 16.50 departure from Leeds and the 18.00 return from Doncaster. It is interesting to note this stopping train consisted of six corridor coaches rather than a two-car DMU.

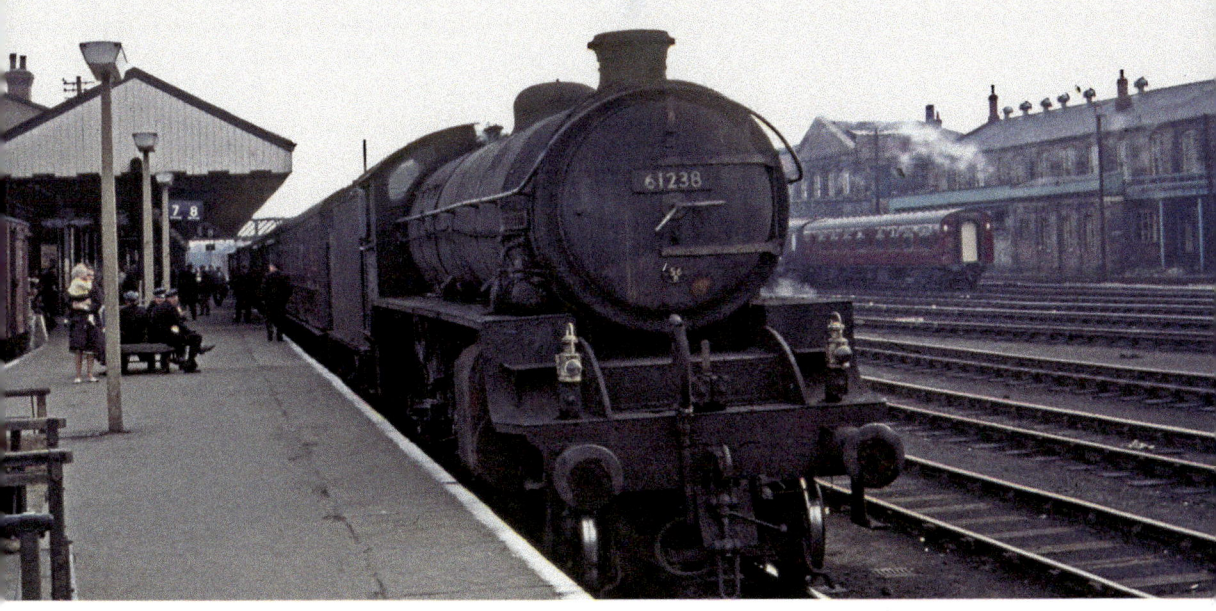

7 June 1965. LNER B1 4-6-0 No. 61238 at Doncaster.

LNER B1 4-6-0 No. 61238 *Leslie Runciman* had been built for the LNER by the North British Locomotive Company in Glasgow and entered service in September 1947. It had been based in the north-east until June 1964, when it was reallocated to Ardsley MPD. Its final home was York from October 1965 until withdrawal in February 1967, after which it returned to the north-east for one last time to be scrapped at Bolckows of North Blyth.

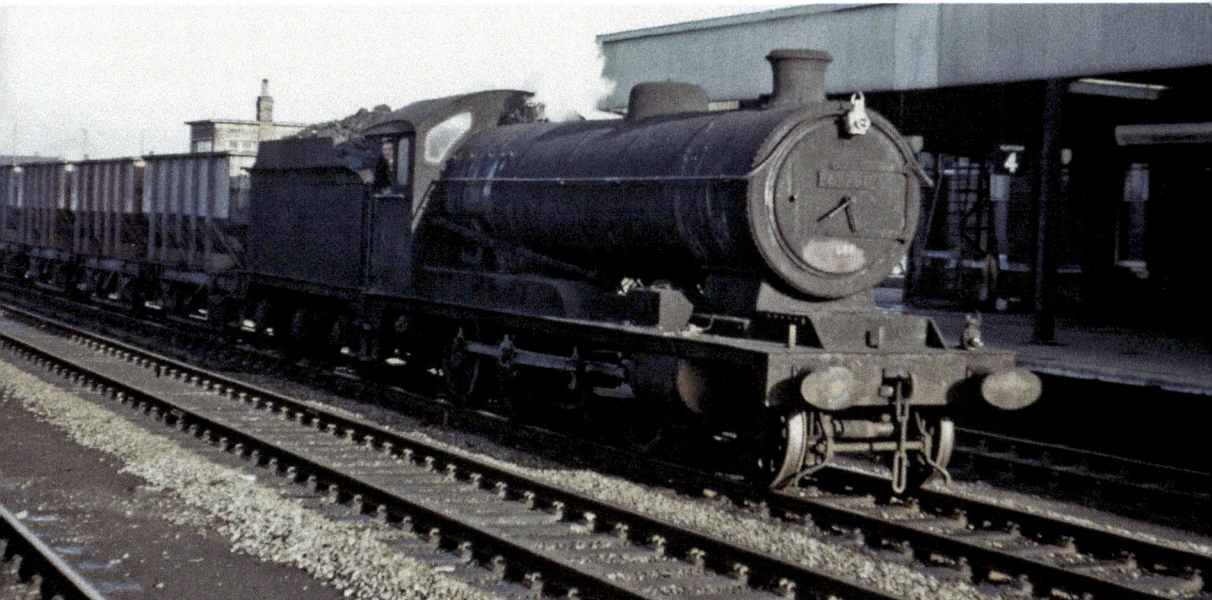

January 1966. LNER O4/8 2-8-0 63781 ambles through Doncaster.

LNER O4/8 2-8-0 No. 63781 was another of the locomotives built for the Railway Operating Division of the Royal Engineers by North British in 1918 and was taken into LNER stock early in 1924. It was rebuilt into an O4/8 in the spring of 1958 and had become a Doncaster-allocated loco with the closure of Colwick depot to steam at the end of 1965. It was to be withdrawn with the end of steam at Doncaster MPD in April 1966.

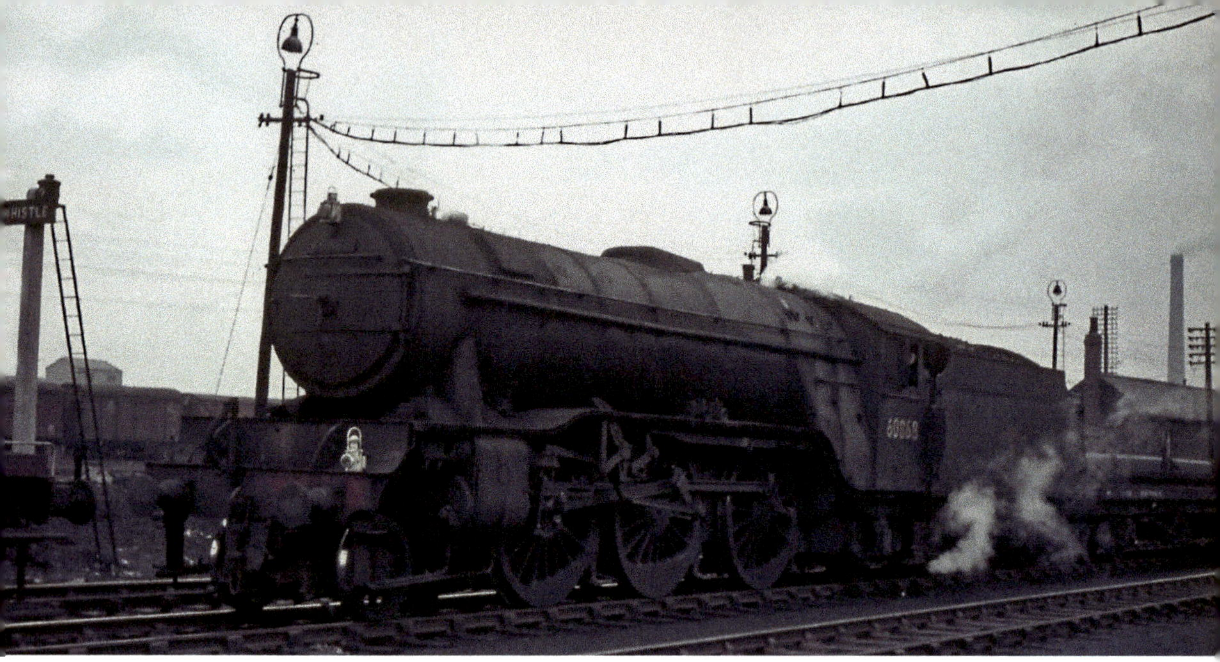

March 1966. LNER V2 2-6-2 No. 60868 waiting at Doncaster.

LNER V2 2-6-2 No. 60868 had been a locomotive allocated to Gateshead and Heaton throughout most of its twenty-seven-year career, until it had been transferred to St Margaret's, Edinburgh, in October 1965. V2s were once an everyday sight in Doncaster but the fourteen remaining at the start of 1966 were rarely seen there. It was welcome, if unexpected, to see this Scottish-based locomotive make an appearance on a steel train. It was in its last six months of service and was withdrawn in September 1966. (L. Flint)

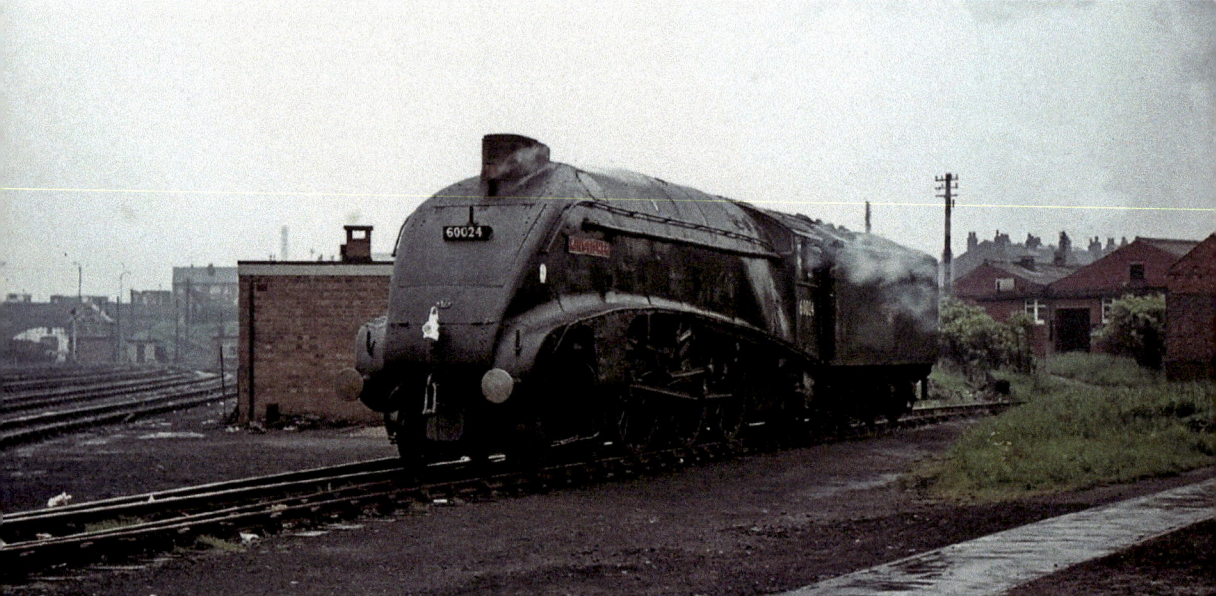

21 May 1966. LNER A4 4-6-2 No. 60024 passes Garden Sidings in Doncaster.

LNER A4 4-6-2 60024 *Kingfisher* prepares to collect the stock for an A4 Preservation Society-organised special, which was called the East Coast Ltd. The train ran between Doncaster and Edinburgh Waverley. It departed at 8.40 a.m. and returned twelve hours later. It is seen passing the site of St James' Bridge station after turning on the triangle of line there. (L. Flint)

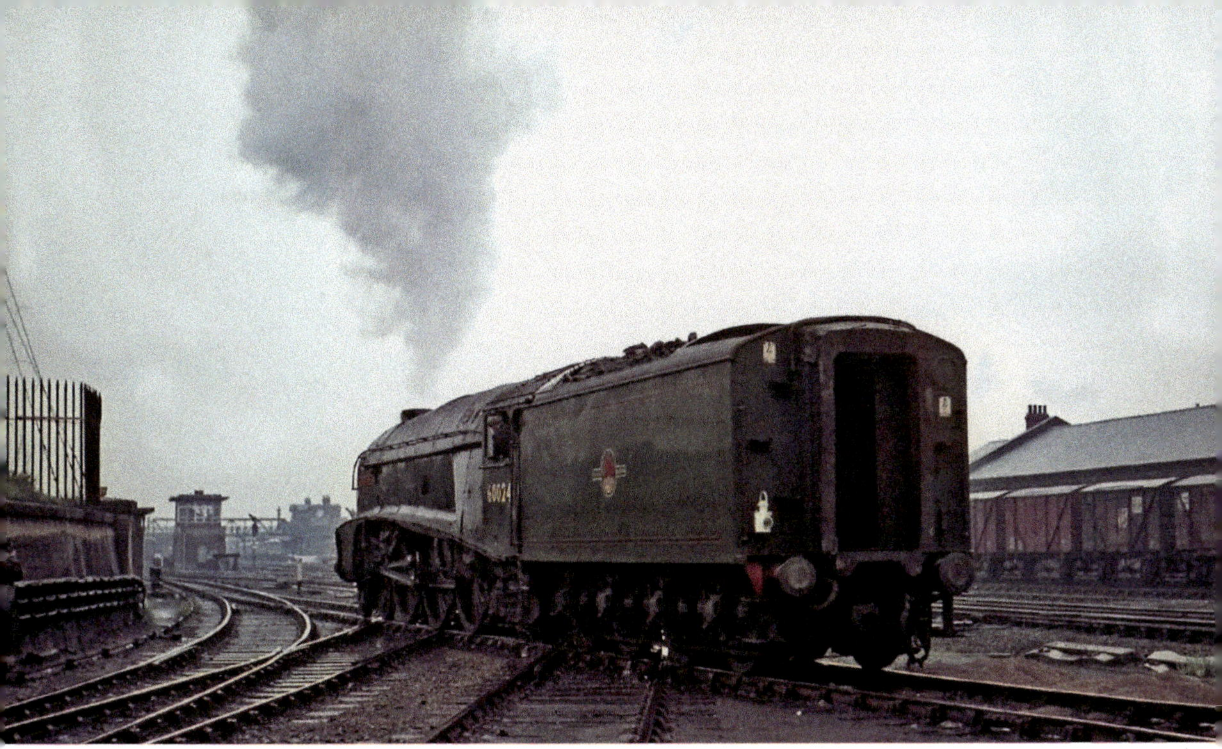

21 May 1966. LNER A4 4-6-2 No. 60024 at Doncaster station.

LNER A4 4-6-2 No. 60024 *Kingfisher* moves towards the station to head the special that was called the East Coast Ltd. The train had been organised by a group hoping to raise enough money to preserve an A4 locomotive. (L. Flint)

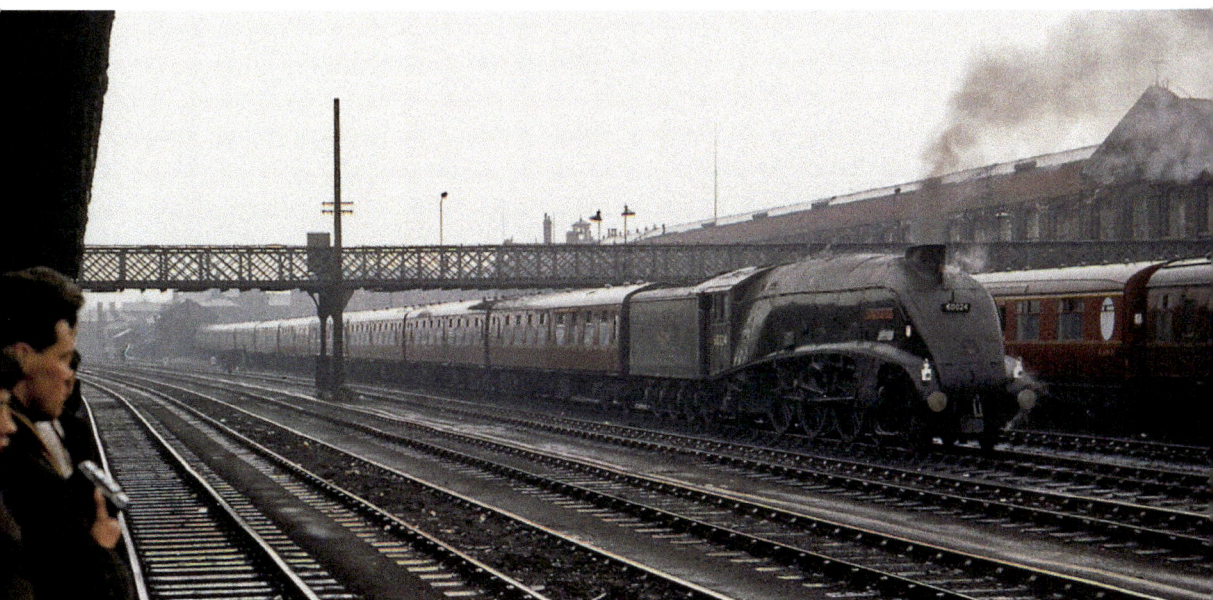

21 May 1966. LNER A4 4-6-2 No. 60024 at Doncaster station.

LNER A4 4-6-2 No. 60024 *Kingfisher* collected its ten-coach special before manoeuvring its train into the platform to load the eager participants. They would enjoy a high-speed run to Edinburgh and back to help the society raise enough money to purchase an A4 locomotive for preservation. Unfortunately, it was not to be No. 60024 that was saved, as it was withdrawn in September 1966 and scrapped by Hughes Bolckows of North Blyth.

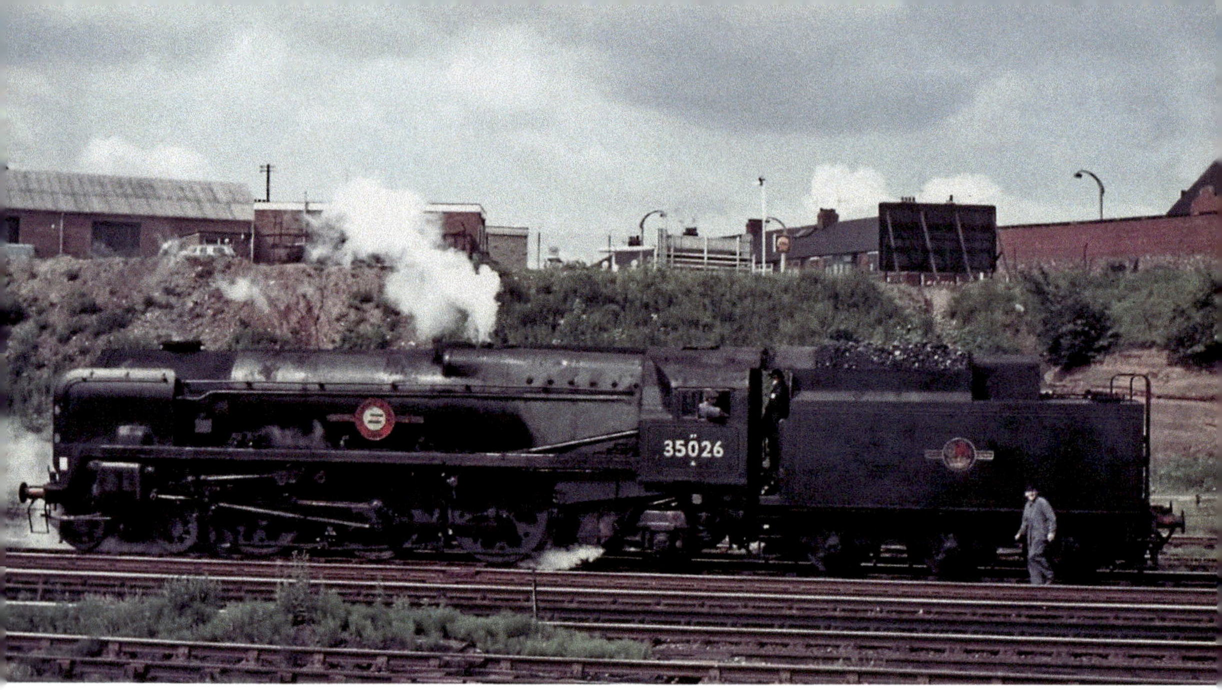

26 June 1966. SR Merchant Navy 4-6-2 No. 35026 makes its way to the depot.

SR Merchant Navy 4-6-2 No. 35026 *Lamport & Holt Line* reverses down to Doncaster station after servicing at the depot. It was to take over the Warwickshire Railway Society rail tour, The Aberdonian. This was three-day steam extravaganza. It involved the use eight locomotives and covered many different railway routes on its way from London Waterloo to Aberdeen and back. (L. Flint)

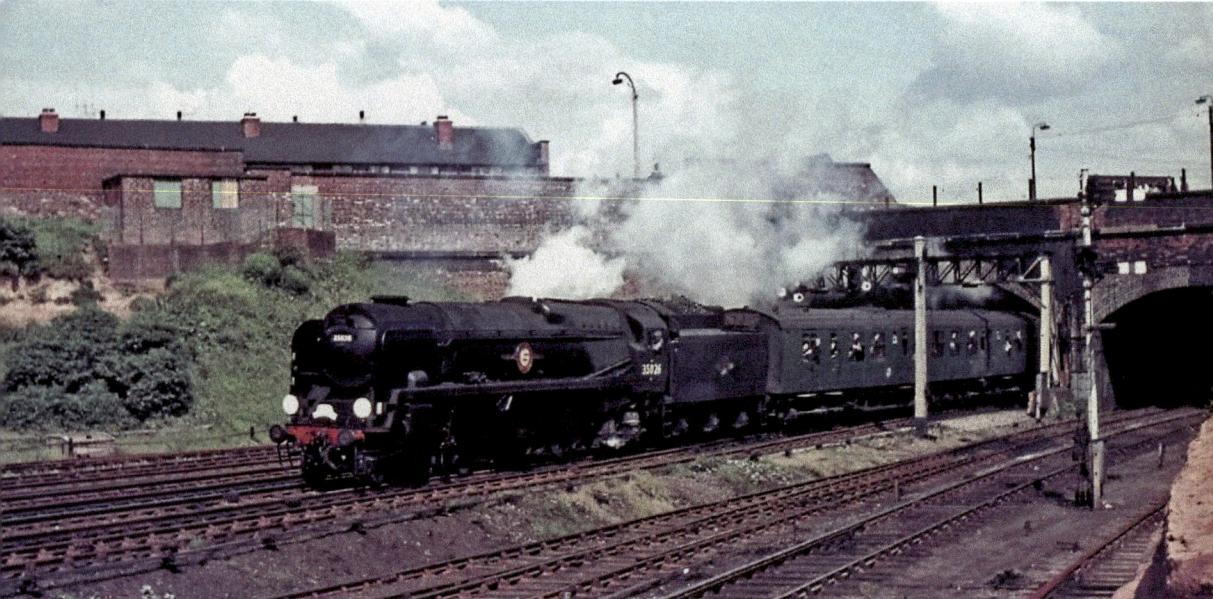

26 June 1966. SR Merchant Navy 4-6-2 No. 35026 departs Doncaster.

SR Merchant Navy 4-6-2 35026 *Lamport & Holt Line* heads southbound from Doncaster. The Warwickshire Railway Society had organised an ambitious three-day rail tour called The Aberdonian. The train began and returned to London Waterloo and traversed many different lines on its circuitous route up to and from Aberdeen. No. 35026 was seen after it had taken over the train at Doncaster on the last leg of the tour. (L. Flint)

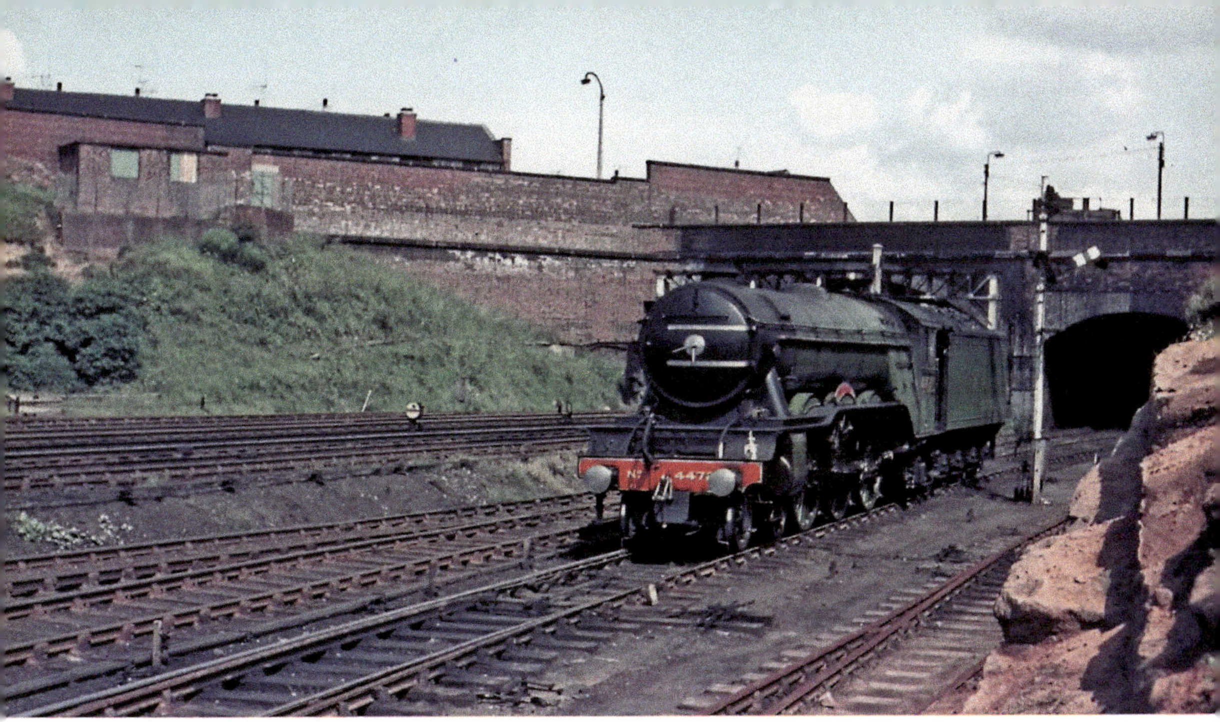

26 June 1966. LNER A3 4-6-2 No. 4472 heads for the depot.

LNER A3 4-6-2 No. 4472 *Flying Scotsman* eases under Balby Bridge, Doncaster, on its way back to the depot. It had worked the Edinburgh to Doncaster part of the return leg of the Warwickshire Railway Society rail tour, The Aberdonian. It was one of the eight locomotives used on the tour.

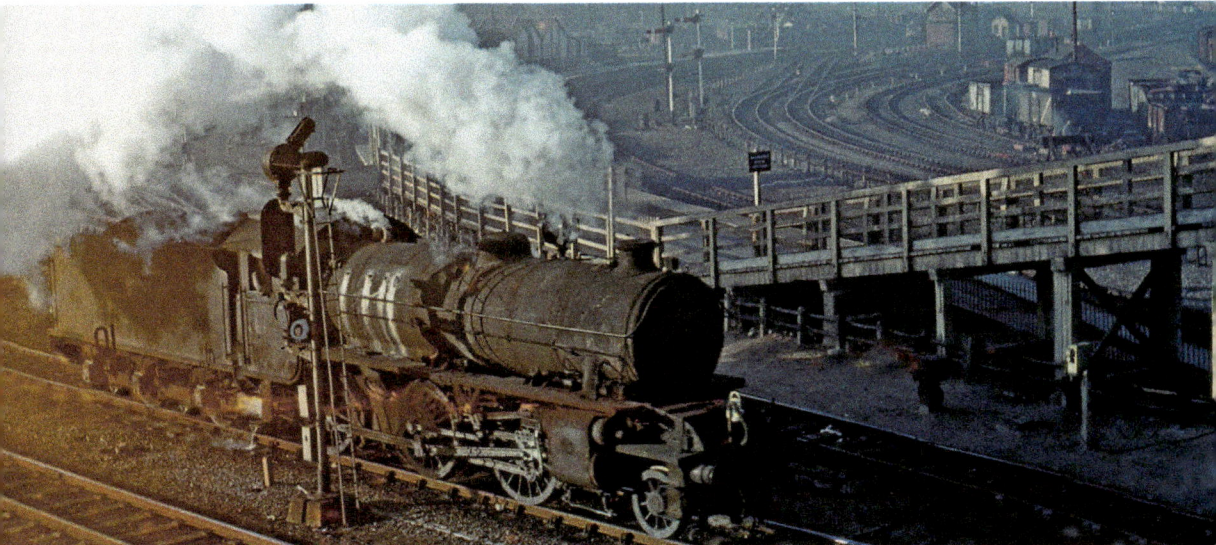

24 October 1966. LNER K1 2-6-0 No. 62028 at St James Bridge, Doncaster.

LNER K1 2-6-0 No. 62028 of York MPD was heading home light engine after turning. By this time Doncaster had ceased to service steam locomotives and visiting engines were quickly sent back out of the area. No. 62028 was to be withdrawn the following month, November 1966. The railway lines curving to the right behind the engine was the route to Sheffield and the curving platform and pedestrian walkway to St James' Bridge was the remains of the excursion used on race days. For years the area was thronged with trainspotters who descended on Doncaster each Saturday. (L. Flint)

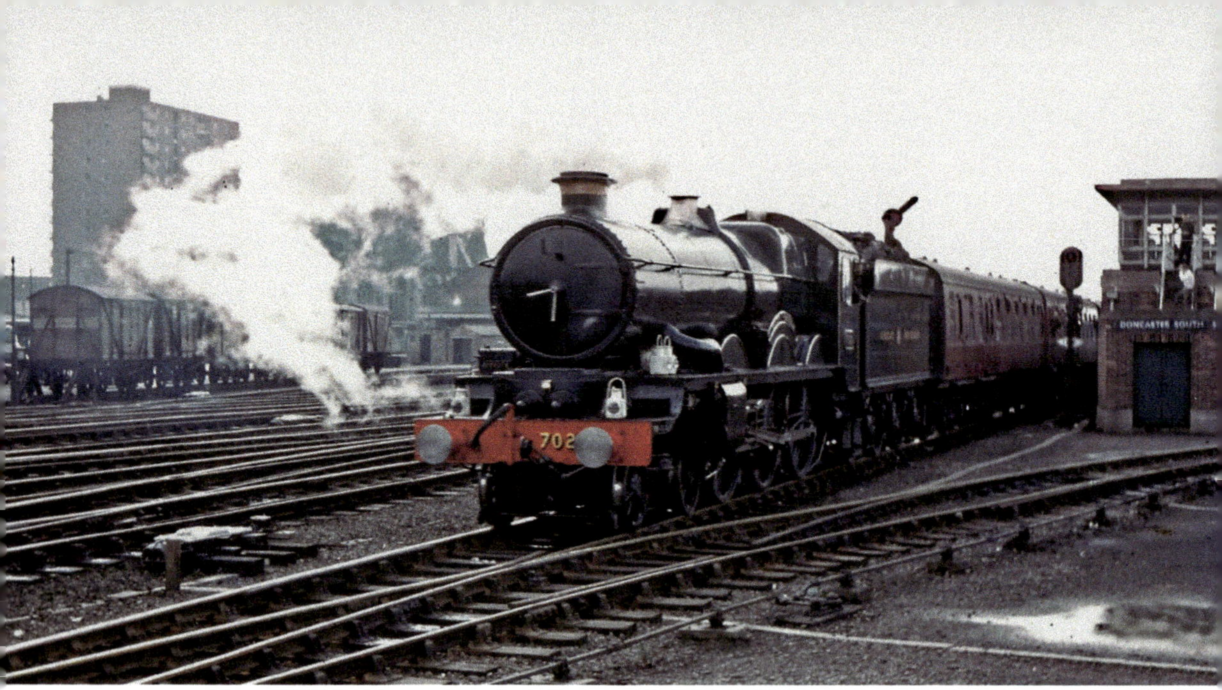

17 September 1967. GWR Castle 4-6-0 No. 7029 at Doncaster.

BR Castle 4-6-0 No. 7029 *Clun Castle* was built at Swindon in May 1950 and after withdrawal in December 1965 from Gloucester Horton Road MPD it was bought for preservation by Patrick Whitehouse. It was seen restarting an Ian Allan special from Kings Cross to Leeds. It was the first visit of the class to Doncaster since the famous trials with No. 4079 *Pendennis Castle* in 1925.

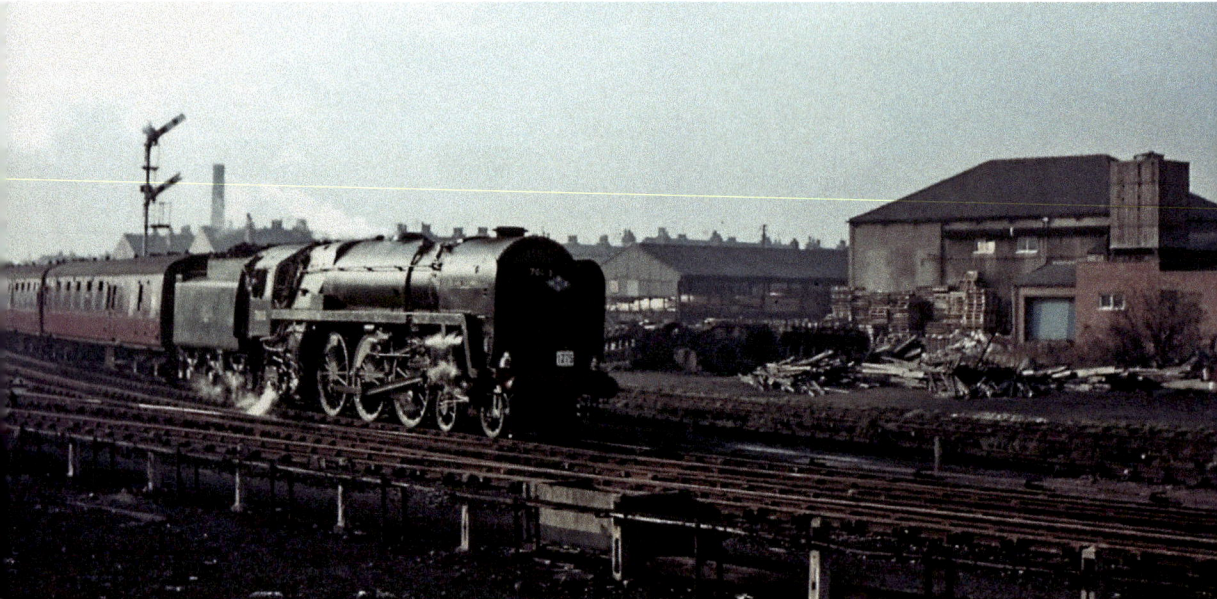

28 October 1967. BR Britannia 4-6-2 No. 70013 heads into Doncaster.

BR Britannia 4-6-2 No. 70013 *Oliver Cromwell* heads a jointly organised Manchester Rail Travel Society and Severn Valley Railway Society Preservation Special into Doncaster along the Sheffield line. The train was hauled by No. 70013 from Manchester Victoria and passed through Doncaster on its way to Normanton where LMS Jubilee 4-6-0 No. 45562 Alberta hauled the train in a circuit around West Yorkshire. (L. Flint)

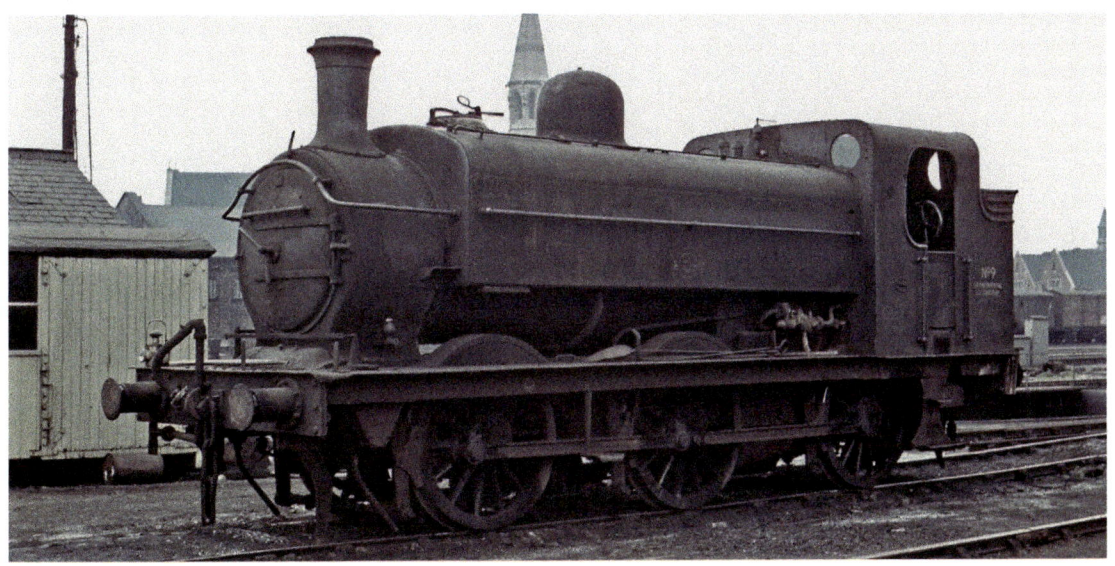

January 1960. LNER J52 0-6-0T Departmental No. 9, ex-No. 68840, at Doncaster.

LNER J52 0-6-0T Departmental No. 9 was designed by Patrick Stirling and had been built in 1899 by Sharp Stewart of Glasgow for the Great Northern Railway as No. 1241. It became British Railways No. 68840 in 1948 and continued to work from its long time home depot of New England. In April 1957 it was transferred to Doncaster and became Departmental No. 9 in January 1958 to work specifically as a shunter in the Plant. It was eventually replaced from these duties by a class J50 in 1961.

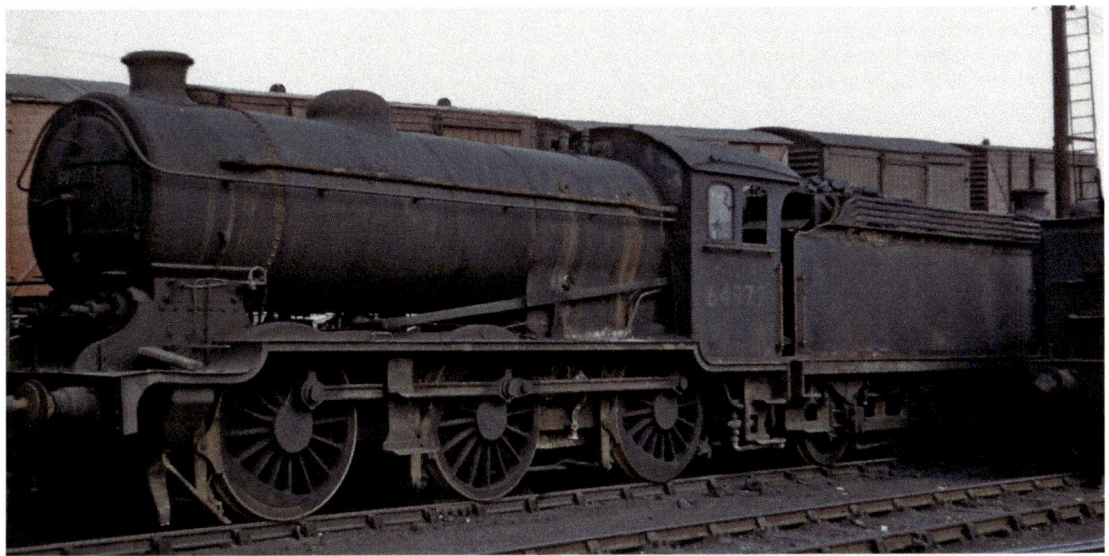

January 1960. LNER J39 0-6-0 No. 64972 awaits entry into the Plant.

LNER J39 0-6-0 No. 64972 was part of the final batch of eighteen locomotives of the class, which had been built at Darlington Works in 1941. It had, when first built, been paired with a second-hand NER tender recovered from a withdrawn locomotive. It had been resident at several Sheffield area depots as well as Ardsley MPD before becoming a Doncaster locomotive in December 1955. It had worked on goods trains and occasional local passenger duties but much of this work had been taken over by the ubiquitous B1s, which seemed to work anything from express passenger services to heavy coal trains.

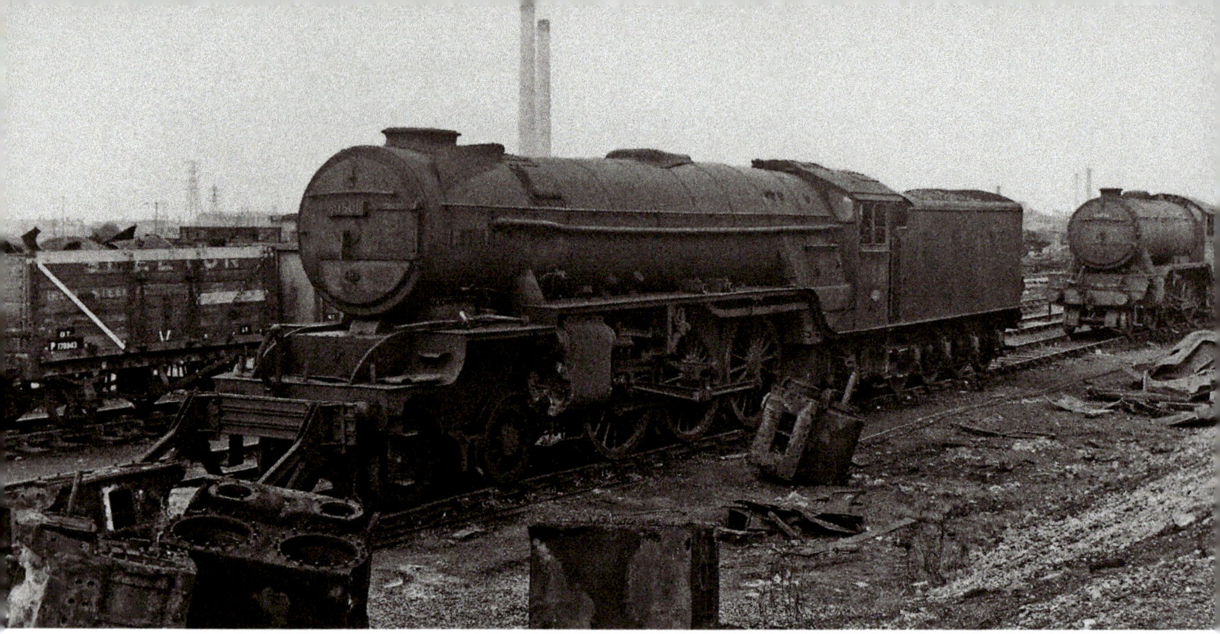

February 1960. LNER A2/2 4-6-2 No. 60501 in Doncaster Plant Works.

LNER A2/2 4-6-2 No. 60501 *Cock of the North* had been built in 1934 as a Gresley P2 2-8-2 in 1934 and by 1938 had been rebuilt with many modifications including Walschaerts valve gear and an A4-style streamlined front end. In 1944 Edward Thompson, the newly appointed Chief Mechanical Engineer, had the locomotive and the other five P2/2s rebuilt into their final form with a 4-6-2 wheel arrangement. No. 60501 had been a York-based engine since 1950 and was withdrawn early in 1960. The A2/2s were the first LNER pacifics to go for scrap in the era of British Railways and for trainspotters, a sudden realisation of what the fate of steam was to be.

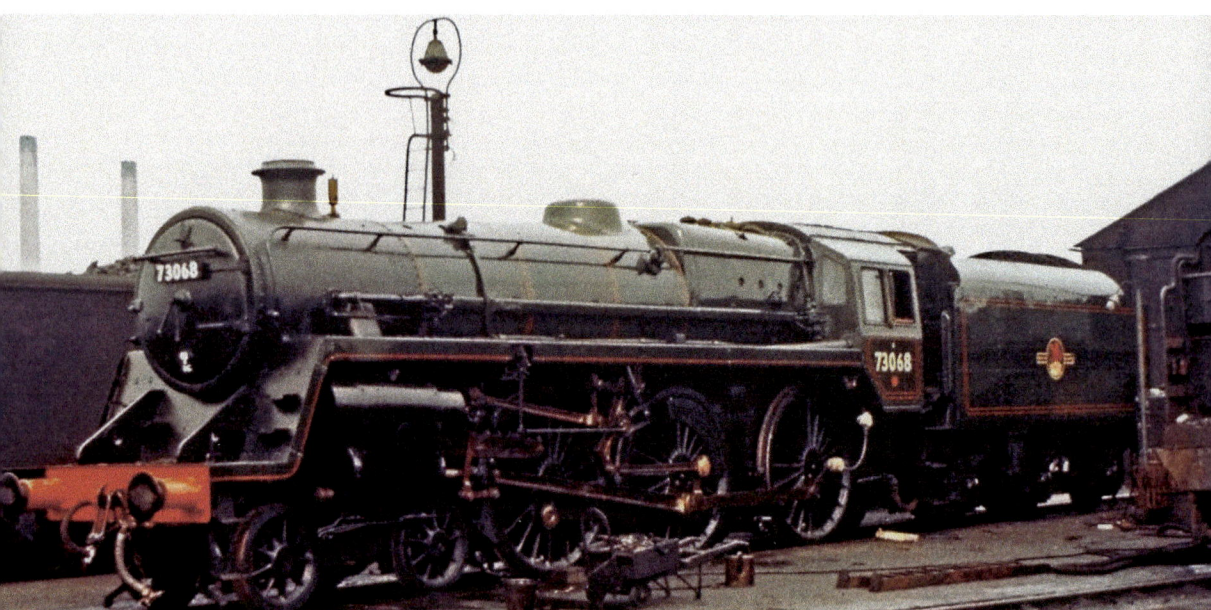

May 1960. BR Standard 5MT 4-6-0 No. 73068 Doncaster Plant.

BR Standard 5MT 4-6-0 No. 73068 looked resplendent in Brunswick green livery after having a general overhaul. It awaits its return to Bristol Barrow Road MPD from Doncaster Plant. No. 73068 had been built at Derby and had entered service in October 1954, becoming a Western Region-allocated locomotive. Its final allocation was to Bath Green Park MPD from where it was withdrawn in December 1965.

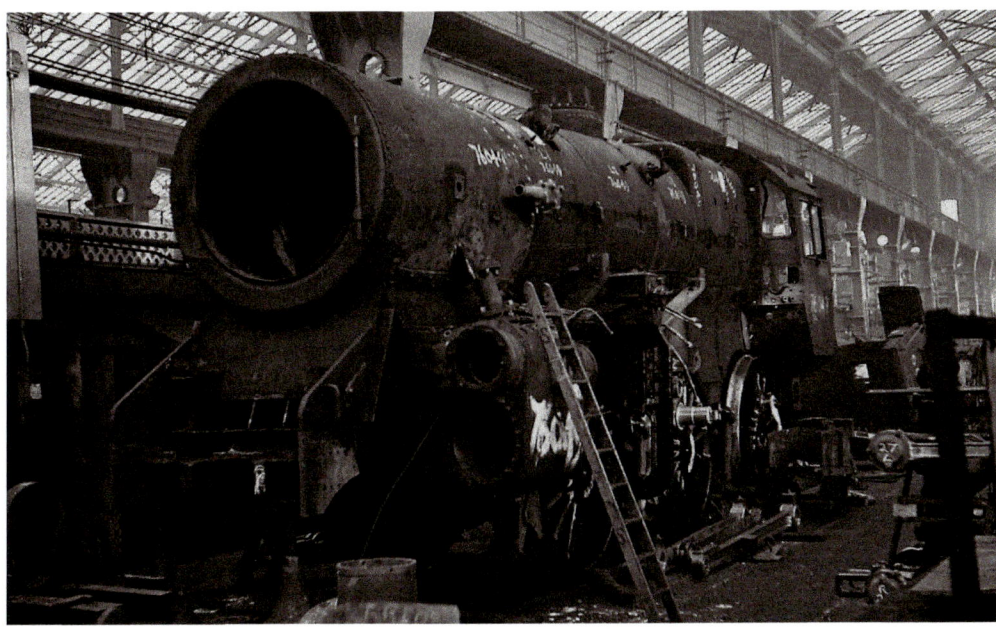

February 1961. BR 4MT 2-6-0 No. 76049 in Doncaster Plant Works.

BR Strandard 4MT 2-6-0 No. 76049 was receiving a general overhaul in Doncaster Plant. It had been built there in 1955 but spent much of its life allocated to depots in the north-east of England. Its final two years of active service were in Scotland and it was withdrawn from Bathgate in January 1966. It was a member of this class, No. 76114, that was to be the last steam locomotive built at Doncaster in October 1957. This brought to an end ninety years of steam engine construction, which totalled over 2,200 locomotives.

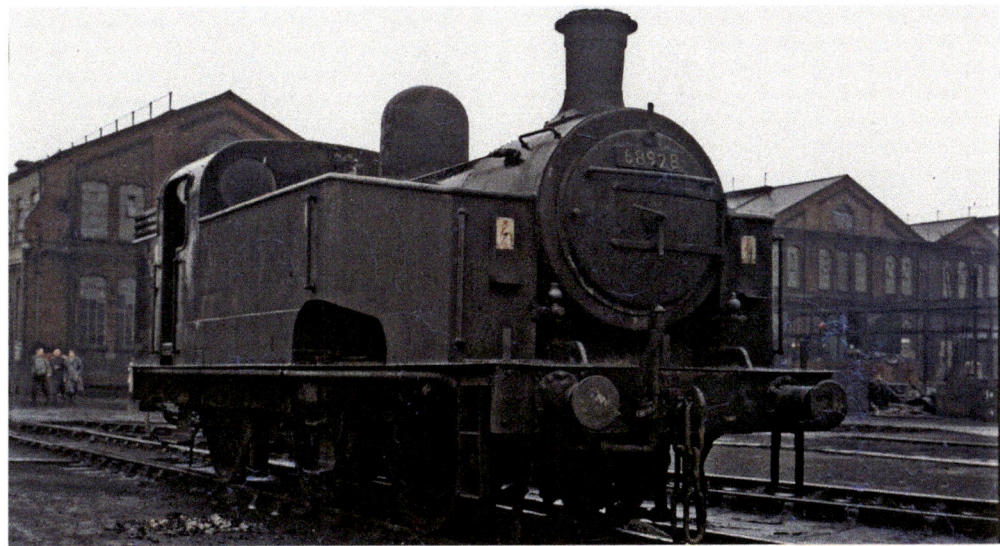

March 1961. LNER J50 0-6-T No. 68928 in Doncaster Plant Works.

LNER J50 0-6-T No. 68928 was seen working as a shunter in the Plant in early 1961. It had recently been reallocated to these duties from Hornsey MPD and was yet to be allocated the Departmental No. 13. It continued in its new role along with several other J50s until they were withdrawn in May 1965. They were all scrapped by T. W. Ward, Beighton, Sheffield.

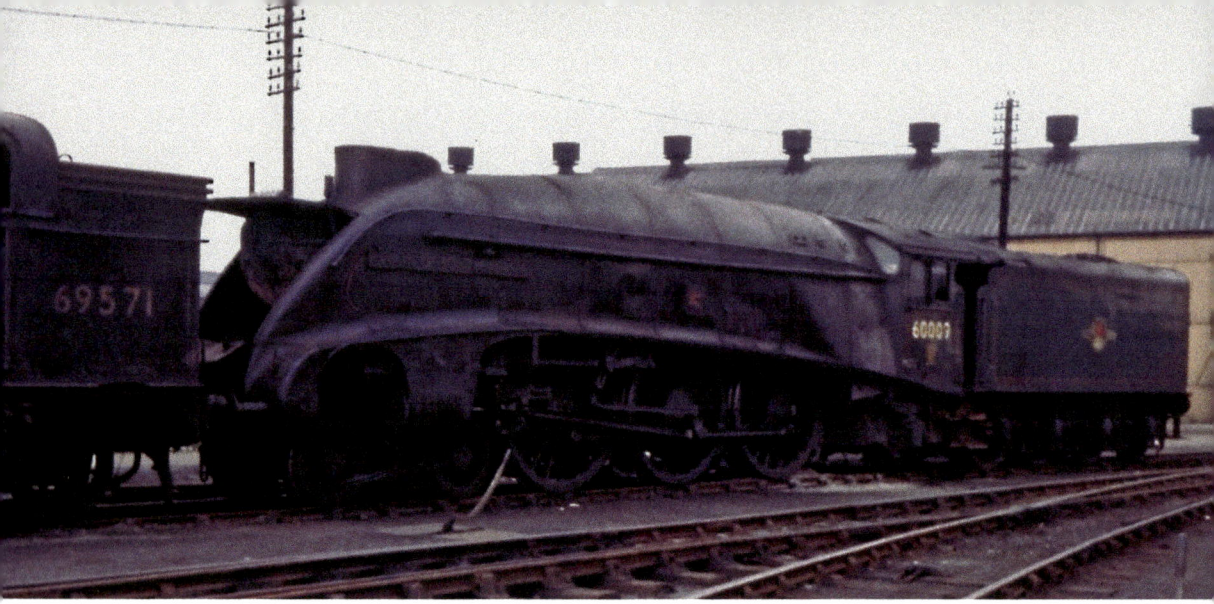

June 1961. LNER A4 4-6-2 No. 60009 in Doncaster Plant Works.

LNER A4 4-6-2 No. 60009 *Union of South Africa* had entered the Plant for a general overhaul that was to involve a seven-week stay there before it was returned to its express passenger duties in Scotland. It became the last steam locomotive to be overhauled at Doncaster when it completed another general overhaul in November 1963. It was withdrawn from Aberdeen Ferryhill in June 1966 and was purchased for preservation in working order within a month. Since 1973 it has regular headed special trains on the national network with its last main line charter being on March 2020. The N2 0-6-2T No. 69571 next to it was not to be so fortunate; it had been withdrawn from New England MPD and was about to be shunted to the scrapping area.

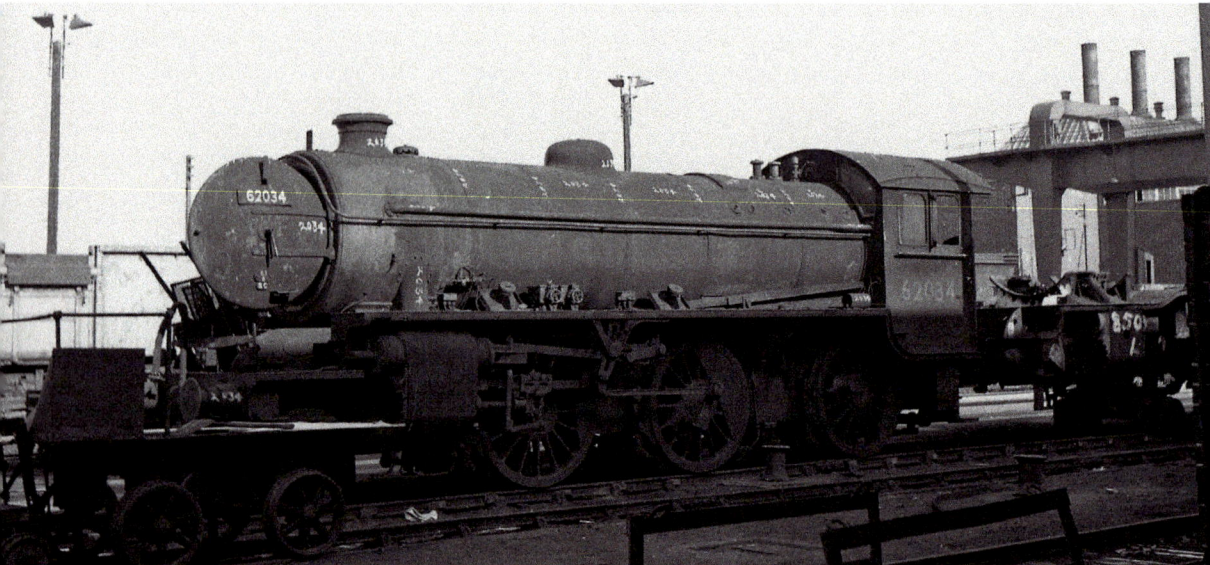

March 1961. LNER K1 2-6-0 No. 62034 in Doncaster Plant Works.

LNER K1 2-6-0 No. 62034 was in the process of been stripped down in preparation for a general overhaul in the Plant Works. Part of that work, the careful labelling of each individual part, can be seen as the locomotive stands outside the erecting shop. The plans for this Fort William-based locomotive were to change, however, and rather than an overhaul to prepare it for a few more years of service it lingered around the Plant for a further eighteen months before it was scrapped.

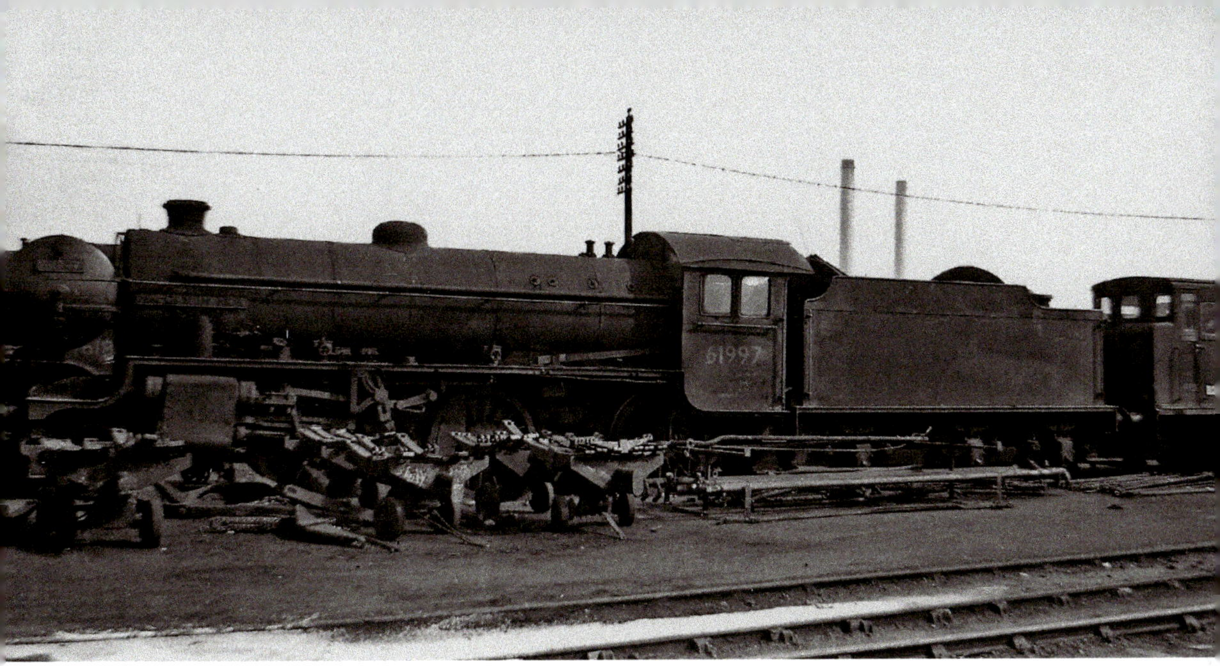

June 1961. LNER K1/1 2-6-0 No. 61997 in Doncaster Plant.

LNER K1/1 2-6-0 No. 61997 *MacCailin Mor* was observed in the rows of condemned locomotives in the Plant having been withdrawn from Fort William MPD. It had been rebuilt in 1945 from a Gresley three-cylinder K4 locomotive into a two-cylinder engine by Arthur Peppercorn, under the direction of Edward Thompson, the then Chief Mechanical Engineer of the LNER. When Peppercorn took over that office the locomotive became the prototype for a new class of seventy K1 2-6-0s, which were delivered after nationalisation.

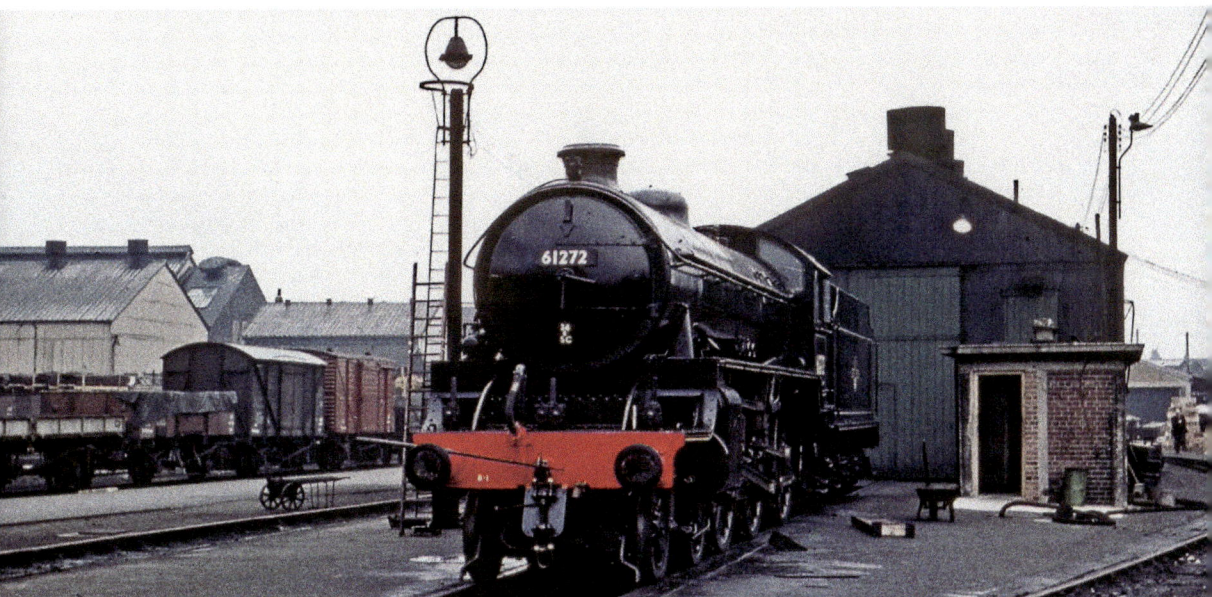

15 September 1961. LNER B1 4-6-0 No. 61272 in Doncaster Plant Works.

LNER B1 4-6-0 No. 61272 was receiving final checks outside the weigh house before being released back into traffic after receiving a general overhaul in the Plant Works. It was to return to its home depot of New England and would remain in regular service until January 1965. It was then transferred to departmental stock, as No. 25, for use as a carriage heating unit. It eventually went for scrap to a private yard in Chesterfield in early 1966.

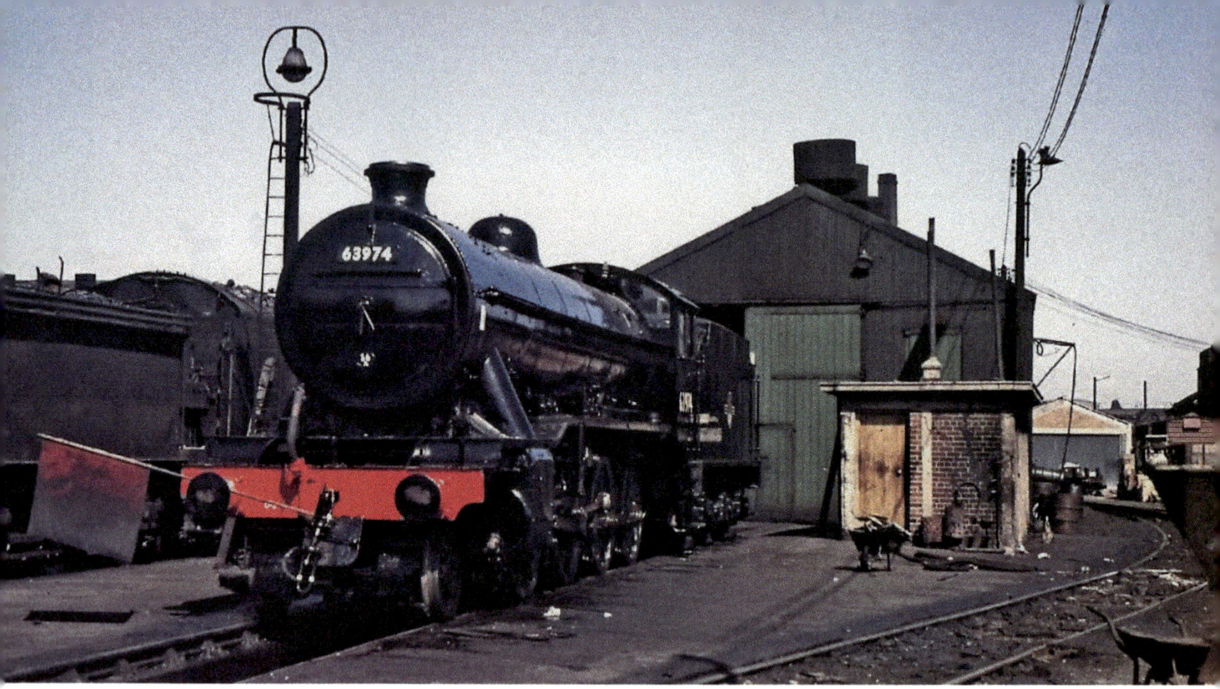

22 April 1962. LNER O2 2-8-0 No. 63974 in Doncaster Plant Works.

LNER O2 2-8-0 No. 63974 stands outside the weigh house for final checks as it nears the end of its last general overhaul. It was to return to Doncaster MPD a few days later to continue its work hauling heavy coal trains. This fine-looking locomotive had less than eighteen months of working life left, however, as dieselisation created a surplus of steam locomotives for the work available.

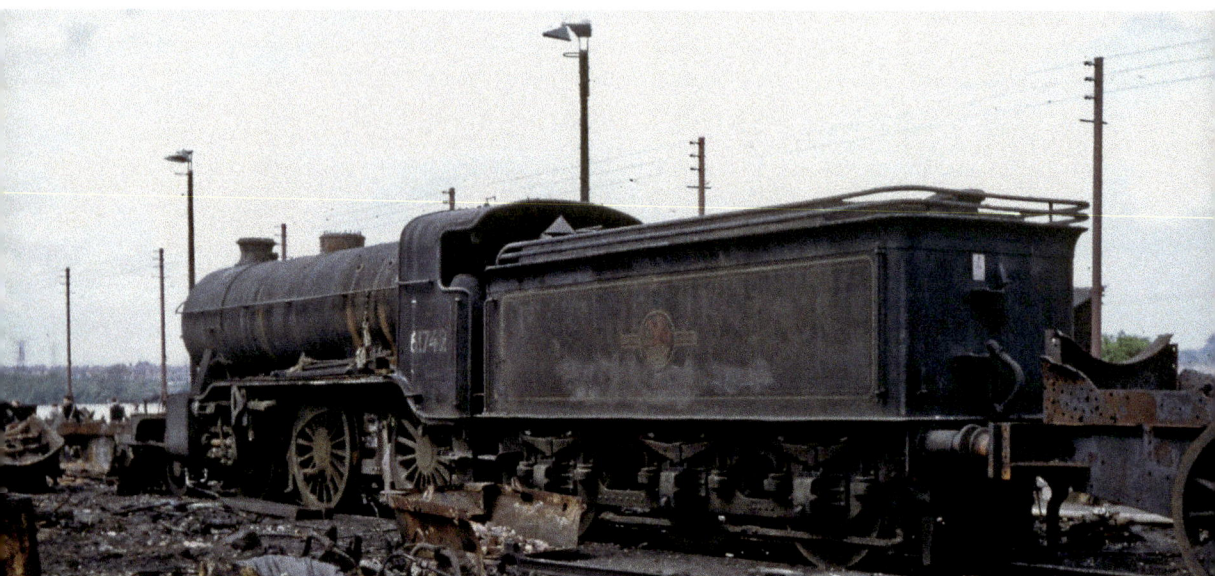

May 1962. LNER K2 2-6-0 No. 61742 Doncaster Plant.

LNER K2 2-6-0 No. 61742 had been designed by H. N. Gresley for the Great Northern Railway and built at Doncaster in May 1916. It had spent its career working from different depots in London and East Anglia before settling for its final years in Lincolnshire. In January 1961 its last reallocation was to New England MPD where it took on the duties as a stationary boiler. By May 1962 it had been dispatched to the Plant for scrapping quickly followed by No. 61756, the last of the class.

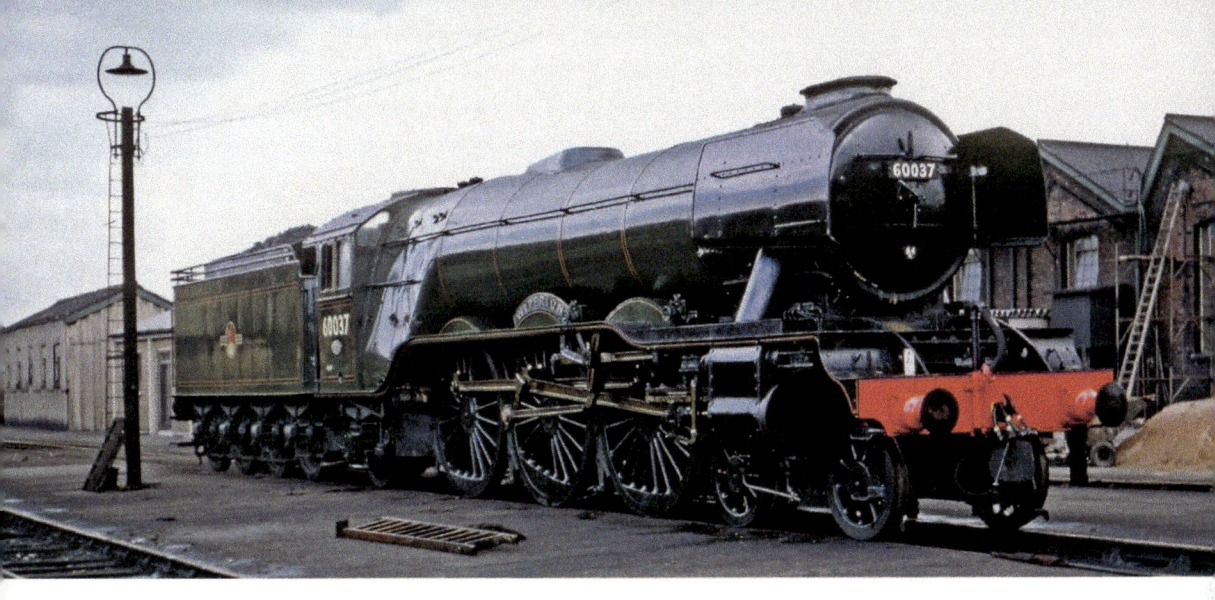

18 May 1962. LNER 4-6-2 A3 No. 60037 at Doncaster Plant.

LNER 4-6-2 A3 No. 60037 *Hyperion* was built at Doncaster in July 1934 and was one of the last batch of A3s to be constructed some eleven years after *Flying Scotsman*. No. 60037 had always been allocated to Scottish depots, either Haymarket or St Margaret's, apart from a month in 1954 when it ventured south to Carlisle Canal. During this visit to the Plant for a general overhaul it was equipped with trough style smoke deflectors. It was finally withdrawn from St Margaret's in December 1964 and scrapped by Arnott Young in Carmyle.

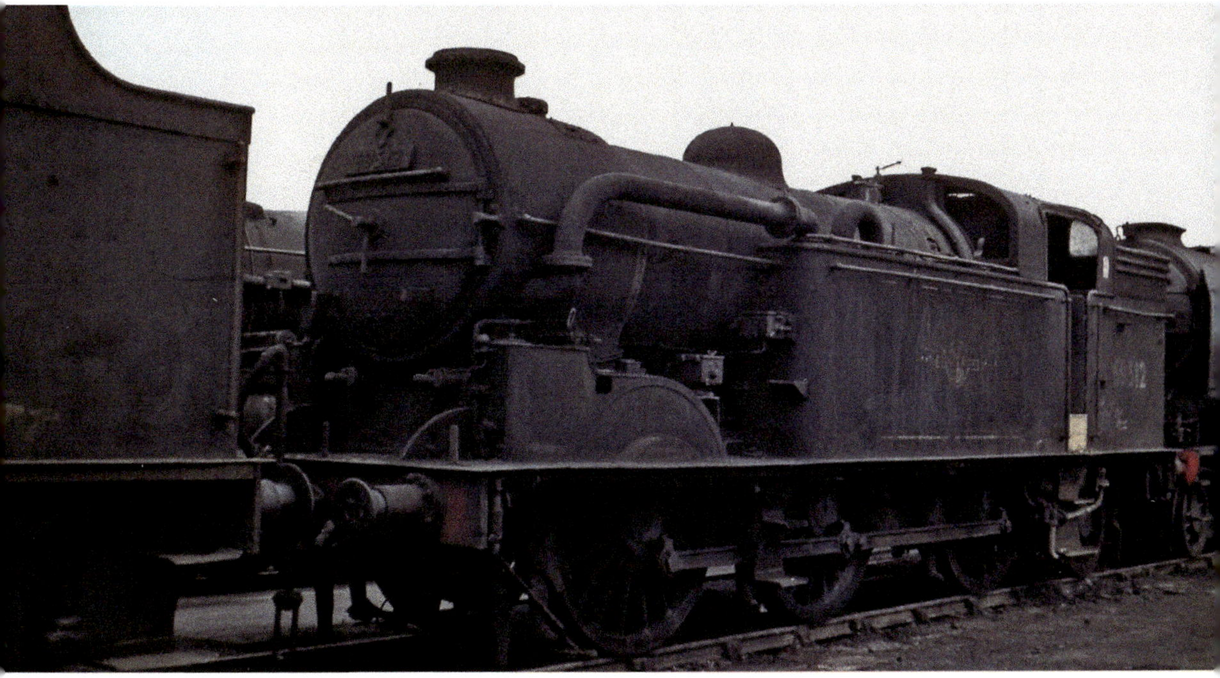

29 September 1962. LNER N2 0-6-2T No. 69512 in Doncaster Plant Works.

LNER N2 0-6-2T No. 69512 had spent most of its forty-one-year career allocated to Kings Cross and Hornsey MPDs working busy commuter services in and out of the capital and it still carried the condensing equipment used on some of those lines. By 1961 it had been replaced by diesel locomotives and had spent the last year of its active life at New England MPD on station pilot's duties before making this final journey to Doncaster.

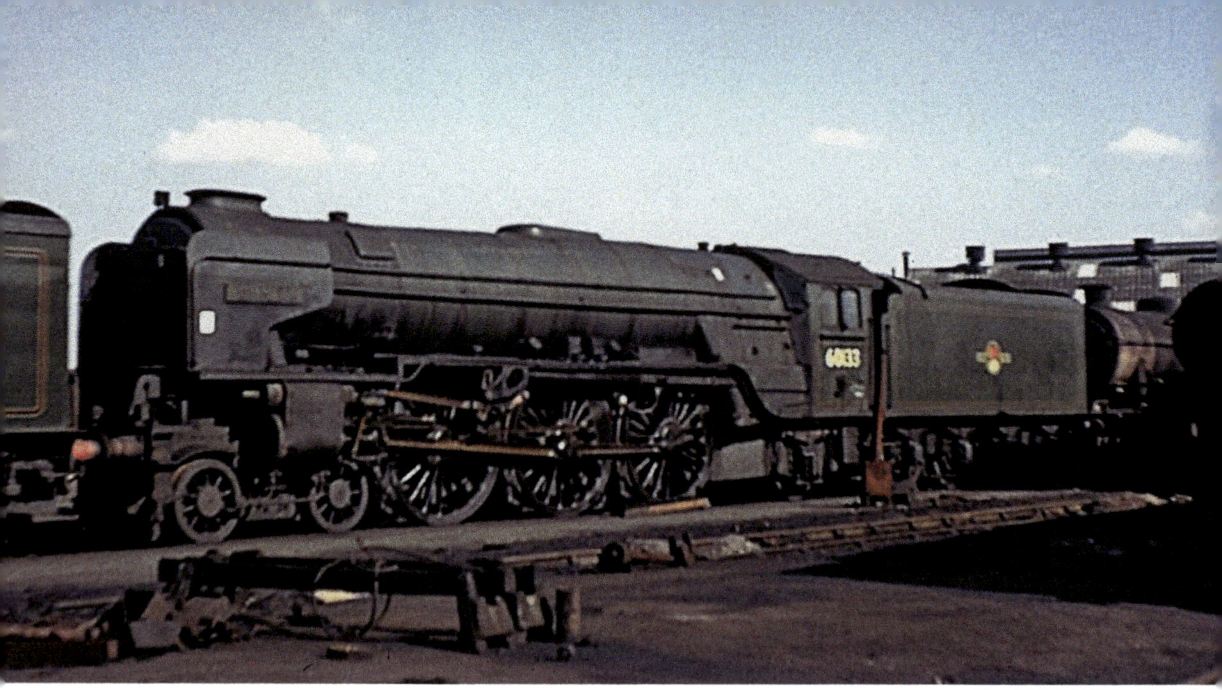

29 September 1962. LNER A1 4-6-2 No. 60133 in Doncaster Plant Works.

LNER A1 4-6-2 No. 60133 *Pommern* was a Peppercorn-designed Pacific built at Darlington Works after nationalisation. It was a Leeds-based engine for most of its life and was to return to Copley Hill MPD after its unclassified repair at the Plant. One final reallocation was to see the locomotive work its last nine months from Ardsley depot and it was finally withdrawn in June 1965.

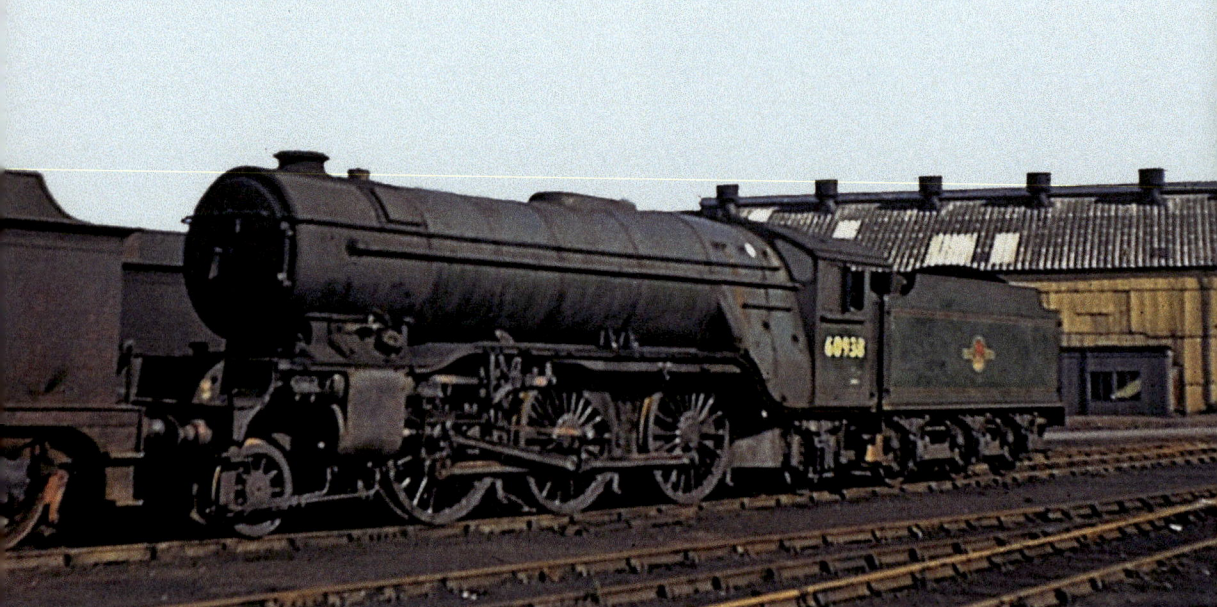

29 September.1962 LNER V2 2-6-2 No. 60938 Doncaster Plant Works.

LNER V2 2-6-2 No. 60938 had been built at Darlington Works in January 1942 and had been allocated to just three depots in its career: Doncaster, March and New England. It had become a Doncaster locomotive six months before this photograph, but with little work for the engine, it was dispatched to the Plant for scrapping.

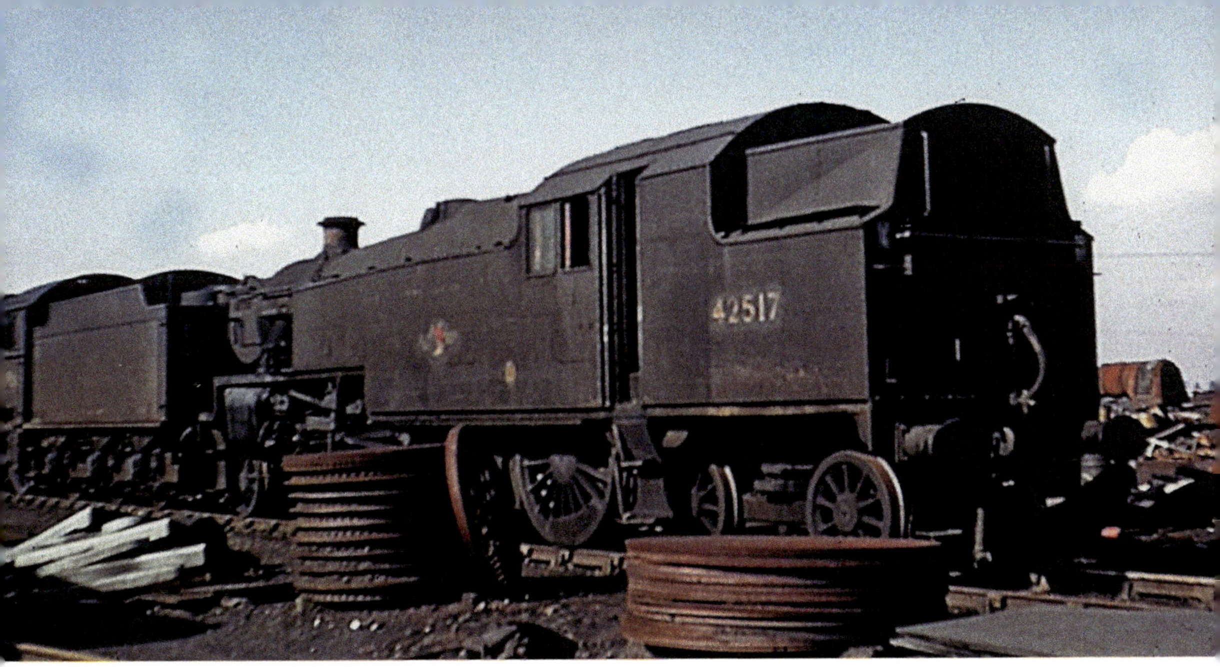

29 September 1962. LMS 4P 2-6-4T No. 42517 in Doncaster Plant Works.

The electrification of the ex-London, Tilbury and Southend Railway between Fenchurch Street and Shoeburyness brought several interesting classes of locomotives to Doncaster for scrapping. Thirty-seven LMS 2-6-4T Stanier three-cylinder locomotives were built at Derby for this heavy outer suburban work, which required rapid acceleration after frequent stops. No. 42517 was one of many of the class to be cut up in the Plant during 1962. No. 42500 was selected for preservation in the National Collection and is on display in the National Railway Museum in York.

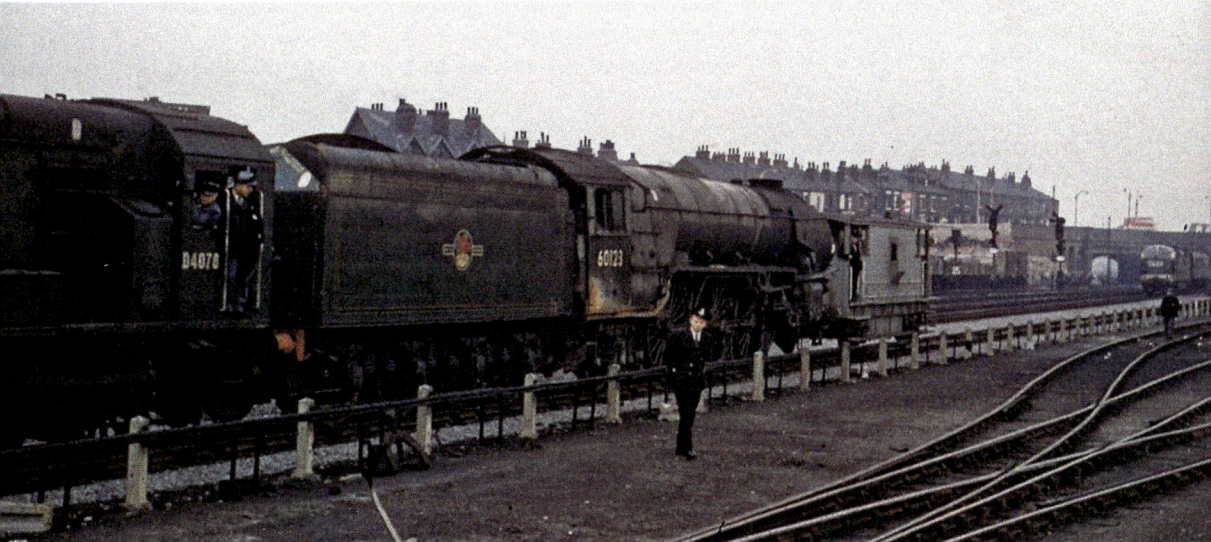

23 September 1962. LNER A1 4-6-2 No. 60123 is hauled to the Plant.

LNER A1 4-6-2 No. 60123 *H. A. Ivatt* entered Doncaster Plant on the 24th September in this condition and was withdrawn on 1 October, the first A1 to be condemned. On 7 September it had suffered severe damage after it derailed at Offord when working the 08.20 Kings Cross goods to Leeds. The front end damage sustained by the locomotive in the accident is clear to see. A few years earlier the locomotive would have been repaired and put back into service rather than consigned to the cutter's torch. The reason for the change of motive power policy looms into sight on the right of the photograph. (Photograph M. Fowler.)

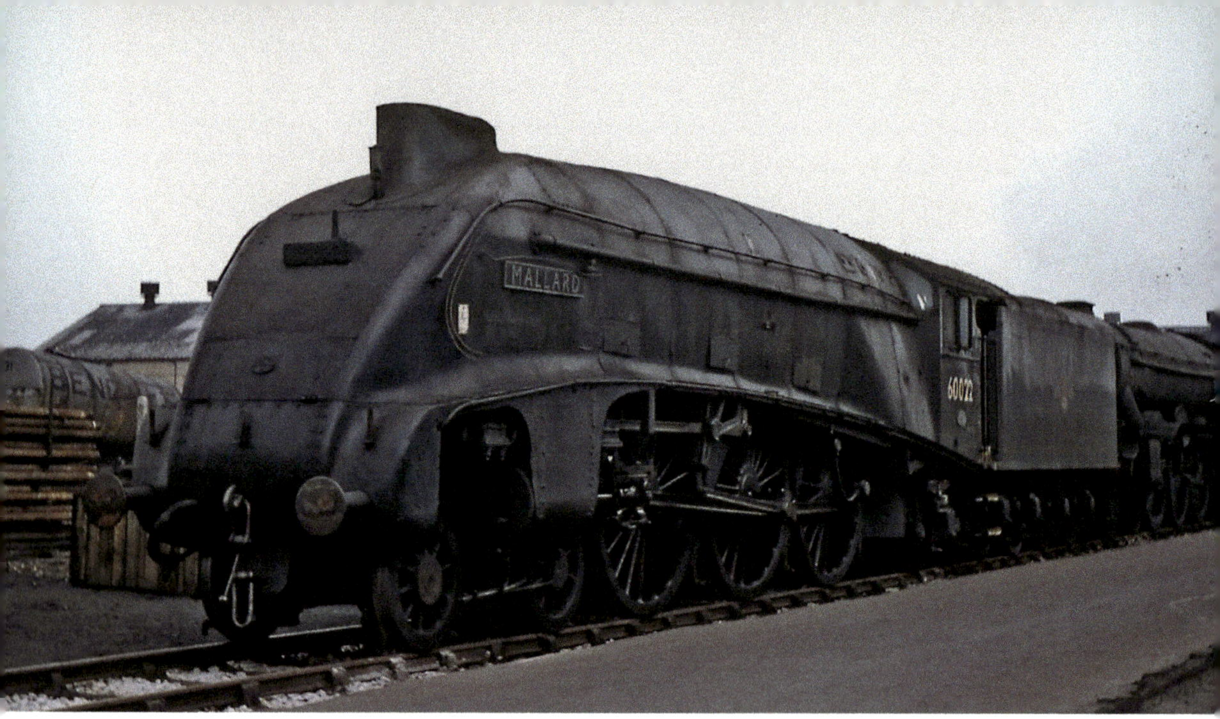

25 November 1962. LNER A4 4-6-2 No. 60022, Doncaster Plant Works.

LNER A4 4-6-2 No. 60022 *Mallard* had arrived at the Plant to have a light casual overhaul, which was to keep it out of traffic until the second week in December. It was to return to Kings Cross MPD on East Coast Main Line duties for another four months before being withdrawn for preservation in the National Collection.

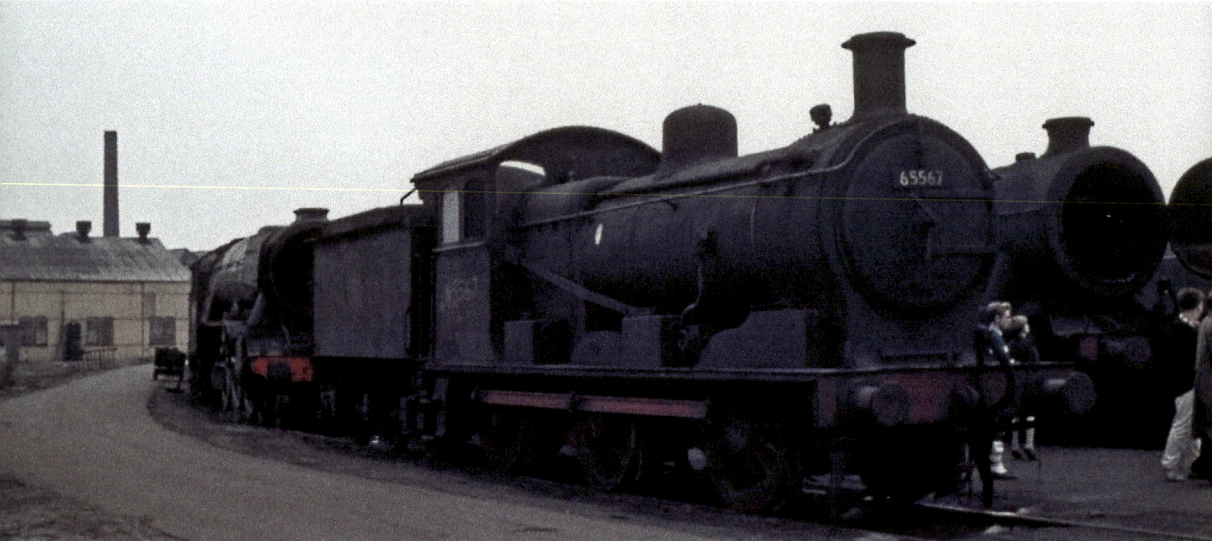

25 November 1962. LNER J17 0-6-0 No. 65567 in Doncaster Plant Works.

LNER J17 0-6-0 No. 65567 had been withdrawn from March MPD three months earlier after a career spanning over fifty-seven years, firstly with the Great Eastern Railway, then the LNER, before joining the ranks of British Railways. It had arrived at the Plant Works, as many other redundant steam locomotives had, but not to be cut up to feed the blast furnaces of Sheffield and Rotherham, but to be preserved. The locomotive had been selected to be part of the National Collection and was cosmetically restored before display at various locations, the most recent being Barrow Hill Roundhouse.

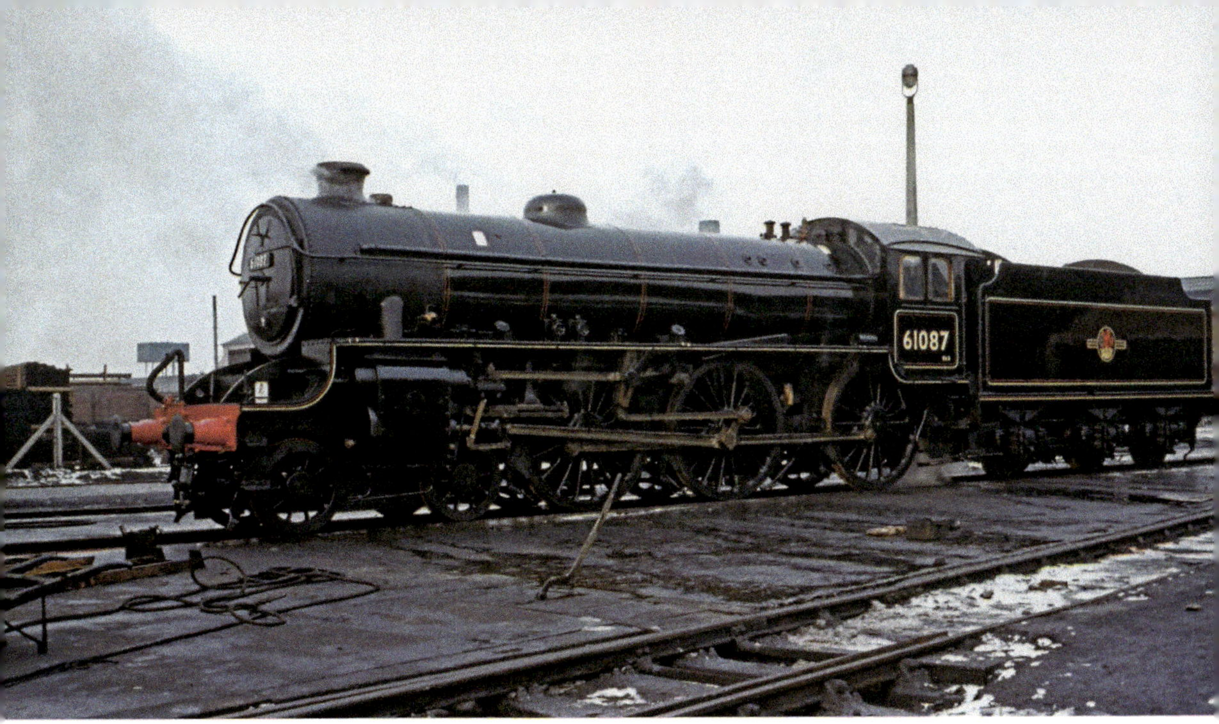

January 1963. LNER B1 4-6-0 No. 61087 in Doncaster Plant Works.

LNER B1 4-6-0 No. 61087 was a long-term resident of Doncaster MPD having been allocated there since February 1949. It looked resplendent in the winter sunshine having just completed a general overhaul in Doncaster Plant. This was to be its last works visit, although it was to remain in service until December 1965.

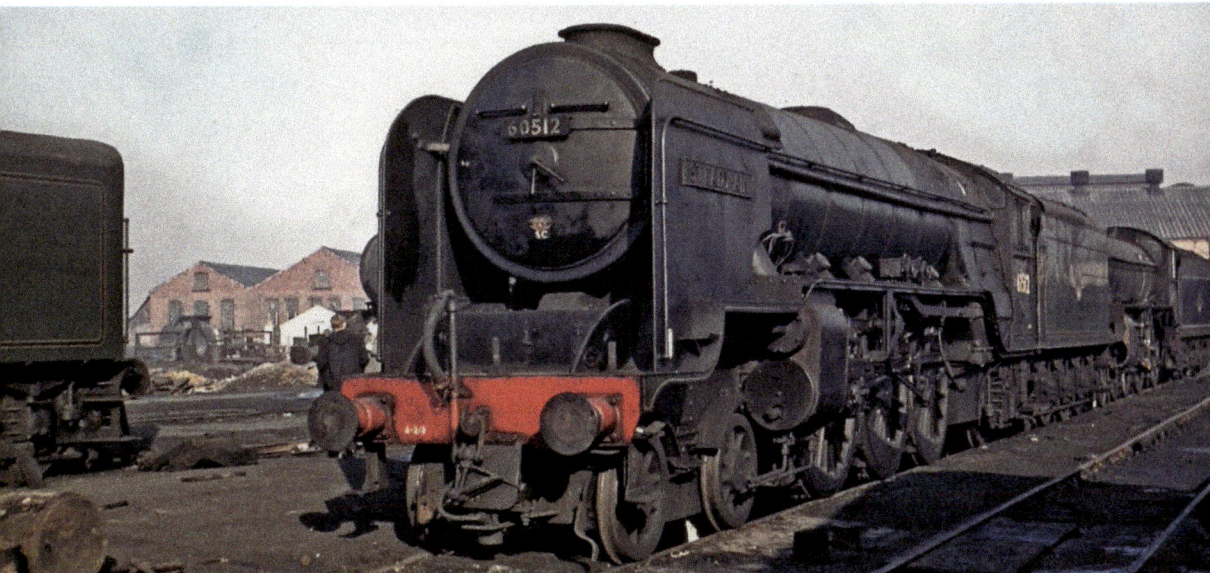

3 March 1963. LNER 4-6-2 A2/3 No. 60512 in Doncaster Plant Works.

LNER 4-6-2 A2/3 No. 60512 *Steady Aim* was built at Doncaster Plant in August 1946 and after working from Gateshead and Heaton depots it became a York locomotive in December 1952. It remained at York until moving north of the border in December 1962 to St Margaret's MPD. By October 1963 it had moved again to Polmadie MPD and within days of its final reallocation to Dundee Tay Bridge it was withdrawn. It was seen in the Plant after arriving for an unclassified repair from St Margaret's.

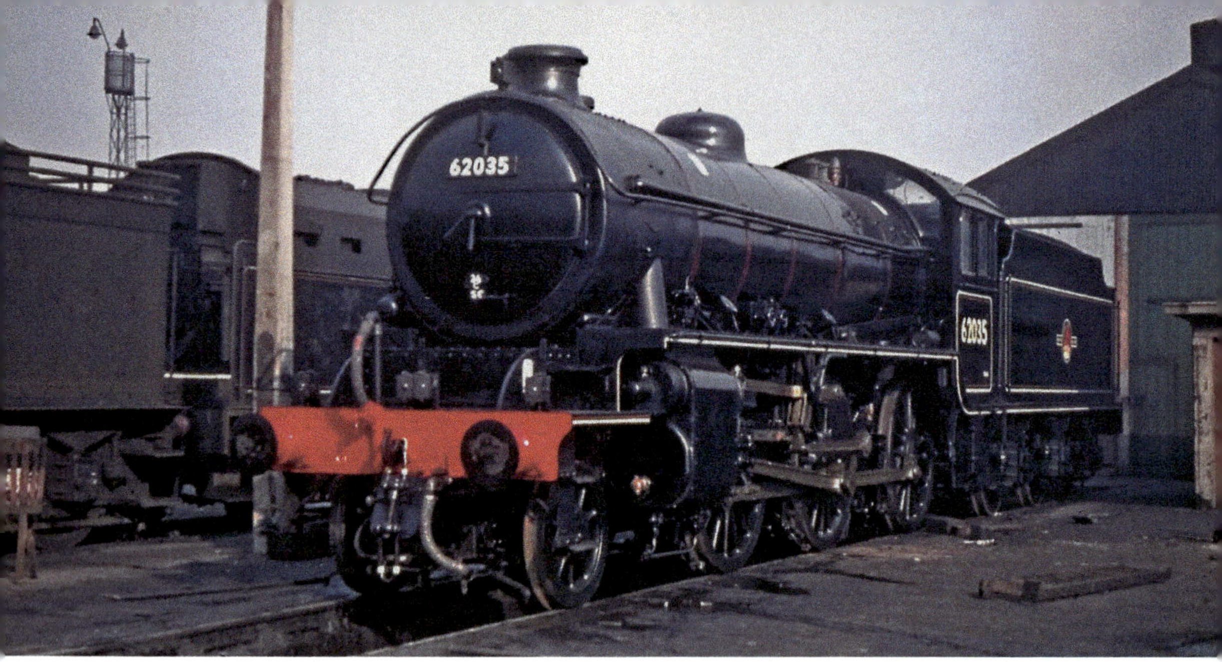

24 February 1963. LNER 2-6-0 K1 No. 62035, Doncaster Plant.

LNER 2-6-0 K1 No. 62035 was seen at Doncaster Plant after its final general overhaul. It was a product of the North British Locomotive Company in Glasgow for the newly formed British Railways and entered traffic in September 1949. It had spent the first ten years of its career based at March MPD where a good number of the class were to be found. No. 62035 was reallocated to Frodingham depot in December 1960 and stayed until withdrawal in July 1965, the year the depot lost several other K1s, which made their way to Draper's of Hull for scrapping.

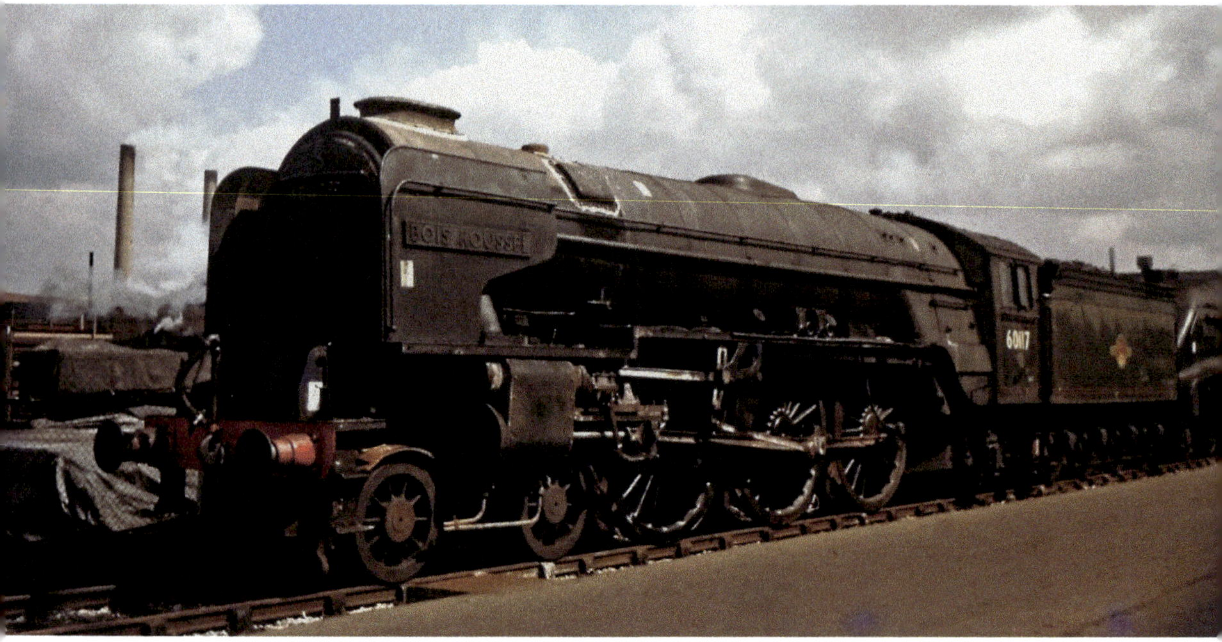

19 May 1963. LNER A1 4-6-2 No. 60117 in Doncaster Plant Works.

LNER A1 4-6-2 No. 60117 *Bois Roussel* in the Plant for its final overhaul, which would allow it to continue in service for another two years. During a shunting movement it had been coupled to an A3 GN coal-rail tender but was reunited with its own tender before being released to traffic a couple of weeks later.

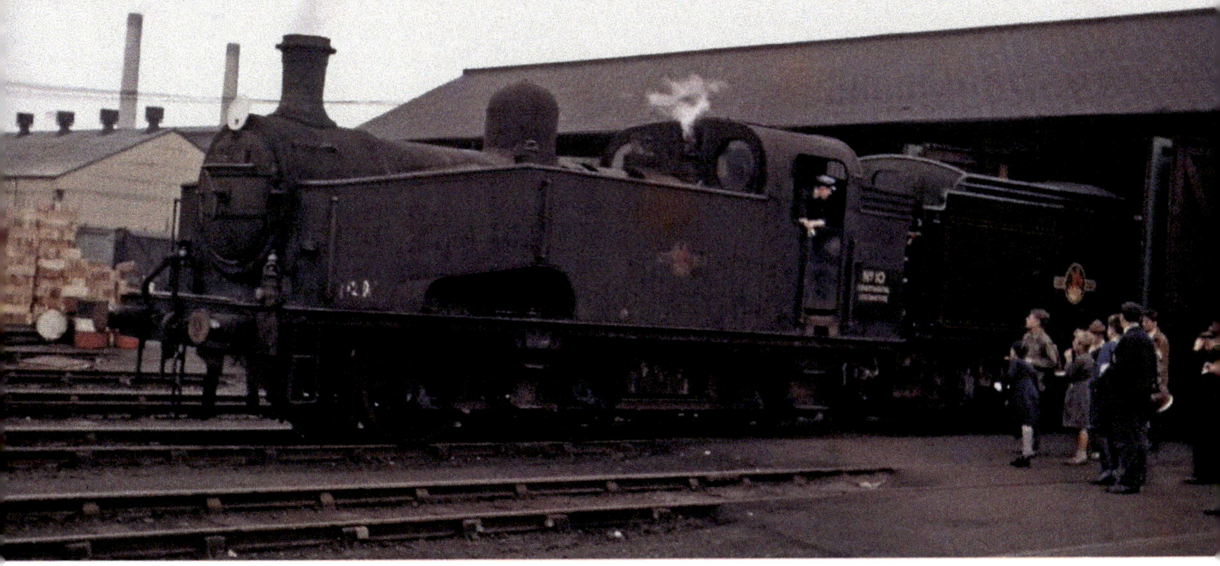

19 May 1963. LNER J50 0-6-0T No. 10 in Doncaster Plant Works.

LNER J50/2 0-6-0T No. 68911 had been built for the Great Northern Railway in 1919 at Doncaster Plant. Very appropriately for an engine belonging to class known as Ardsley tanks, it spent much of its career working from that depot and around Leeds generally. It was withdrawn from normal service in November 1960 and was selected as a works shunter and renumbered Departmental No. 10 in early 1961. It was watched by a group of admiring trainspotters as it manoeuvres an A3 coal-rail tender out of the paint shop to be reunited with the locomotive.

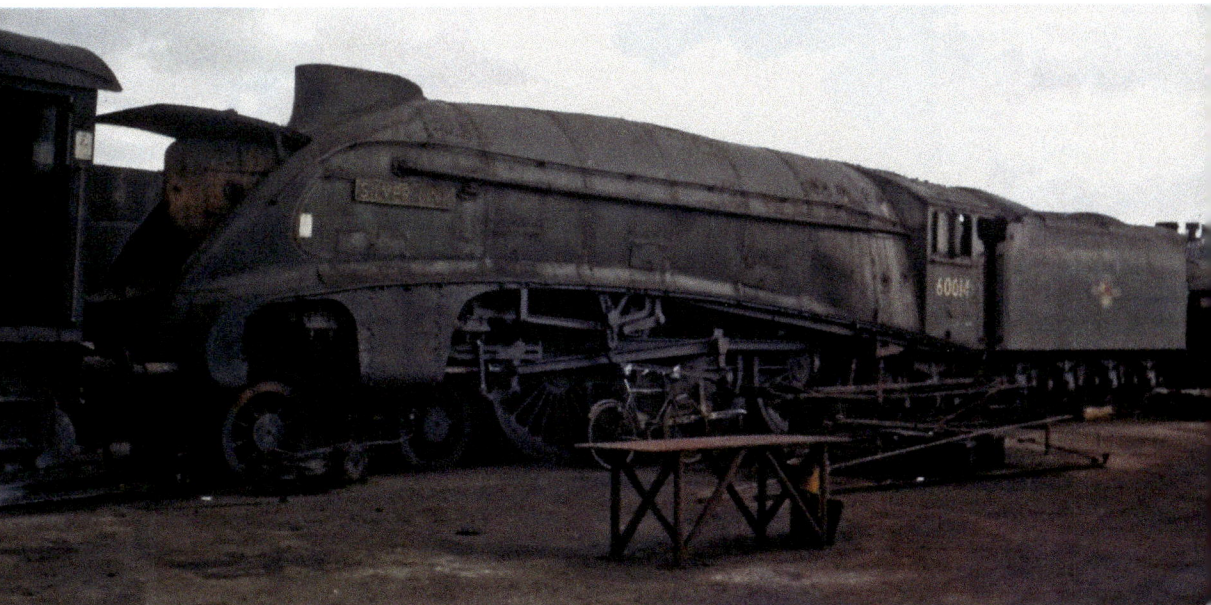

19 May 1963. LNER A4 4-6-2 No. 60014 in Doncaster Plant Works.

LNER A4 4-6-2 No. 60014 *Silver Link* had been withdrawn from Kings Cross shed in December 1962 and when it arrived at the Plant in early 1963, rumours circulated that it was to be preserved by Billy Butlin at one of the holiday camps. As the first of the class it would have been very a fitting retirement for the locomotive, but it was not to be, and No. 60014 was fairly quickly reduced to wagon loads of scrap metal.

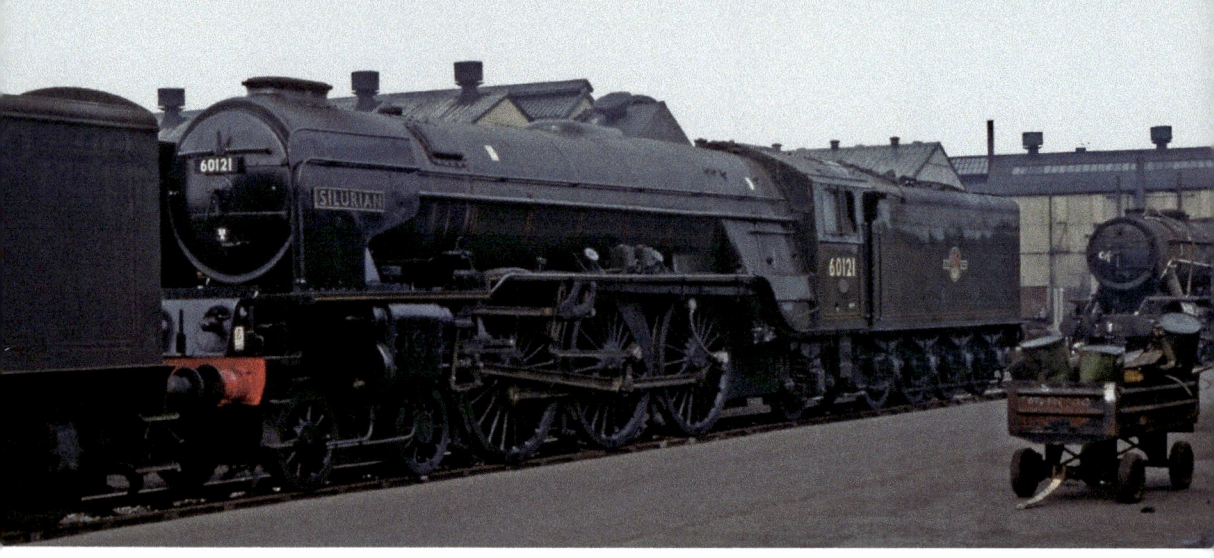

9 June 1963. LNER A1 4-6-2 No. 60121 in Doncaster Plant Works.

LNER A1 4-6-2 No. 60121 *Silurian* was seen outside the Plant during one of the many organised visits there by groups of excited trainspotters. The locomotive spent a total of ninety-seven days in the Plant in 1963 on four separate occasions. It had a general overhaul earlier in the year and had returned later for some problem which had occurred. It was seen here on its third visit for more work to be done. On each occasion it returned to York, the only depot it had been allocated to in its all-too-short career. It would be back in the Plant through most of July having a light casual overhaul which presumably ensured the locomotive was mechanically fit for its final two years in service. It was withdrawn from York in October 1965.

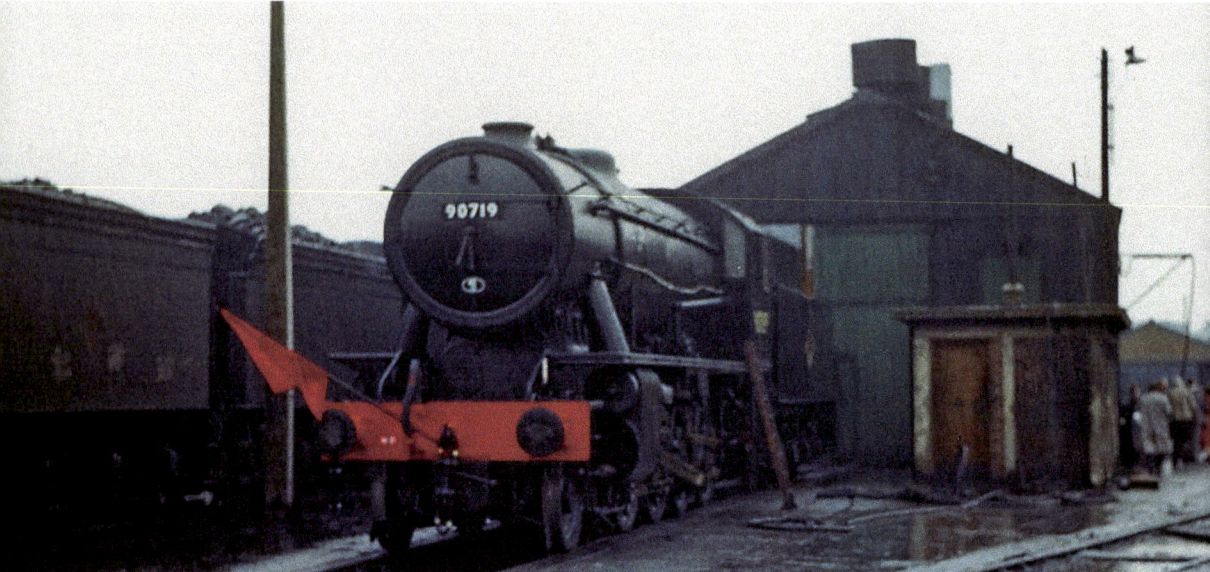

July 1963. WD 2-8-0 No. 90719 in Doncaster Plant Works.

WD 2-8-0 No. 90719 had just completed a general overhaul and was undergoing final checks in front of the weigh house before being released back into traffic. What had been a beautiful summer's day was shattered by a sudden downpour, somewhat spoiling this locomotive's moment of glory. It was to remain at Canklow MPD for nearly two more years before eventually being withdrawn from Langwith Junction in February 1966. It was to be one of the 202 WD 2-8-0s that were cut up by Albert Draper and Son of Hull.

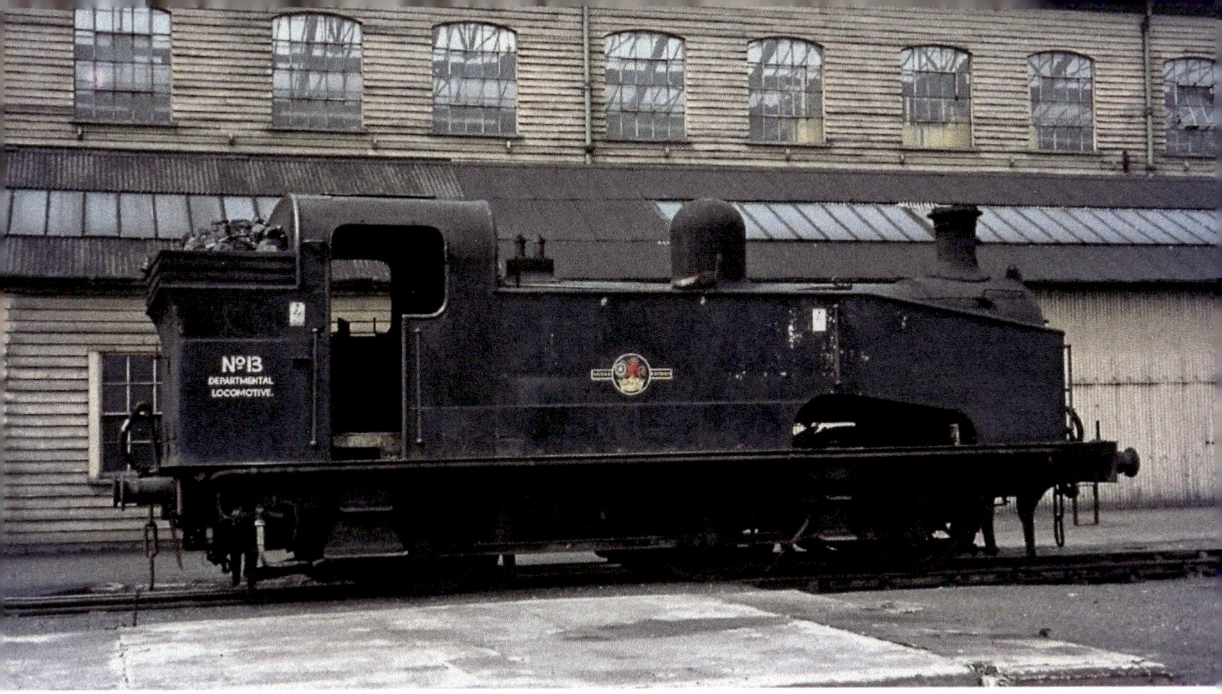

May 1964. LNER J50 0-6-0T Departmental No. 13 in Doncaster Plant Works.

LNER J50 0-6-0T Departmental No. 13 (No. 68928) had been built at the Plant in late 1922 a few months before the Great Northern Railway had been absorbed into the LNER. After nationalisation, the engine had spent almost a decade working around the capital from its Hornsey depot. In 1961 it returned to Doncaster to become a departmental locomotive shunting at the Plant.

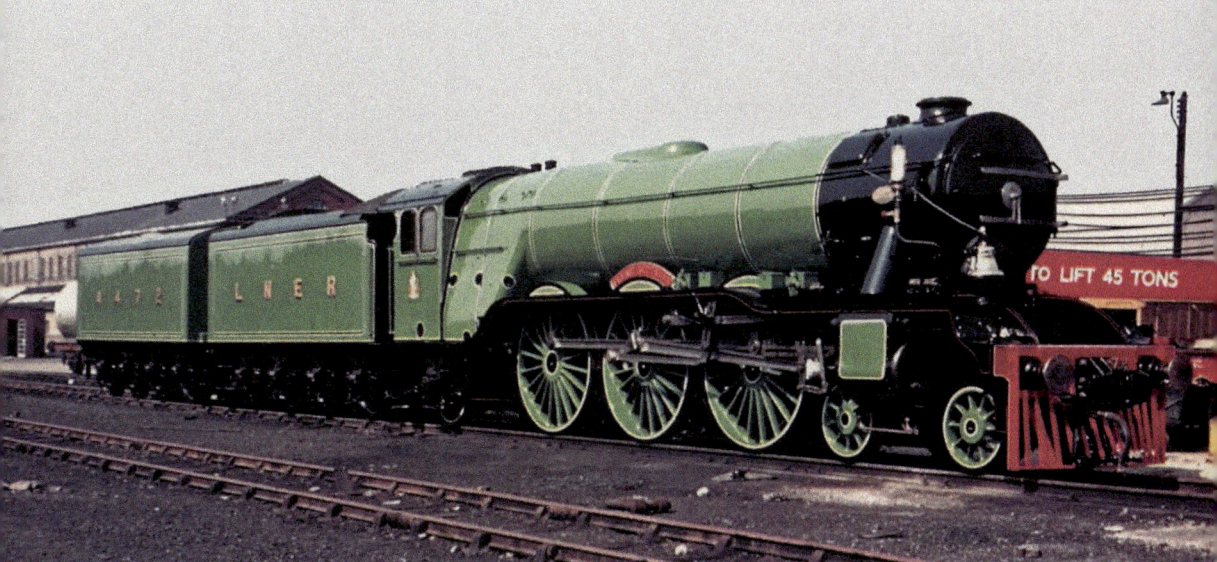

September 1969. LNER A3 4-6-2 No. 4472 in the Plant works.

LNER A3 4-6-2 No. 4472 *Flying Scotsman* was modified and prepared for its North American tour at the Plant. *Flying Scotsman* left Liverpool on 19 September 1969 and steamed over 15,400 miles, taking it across much of the United States of America and even as far as Canada. The tour had the backing of the government and was a mission of goodwill to promote trade and co-operation between British and American businesses with attention drawn to British exports to the USA. (L. Flint)

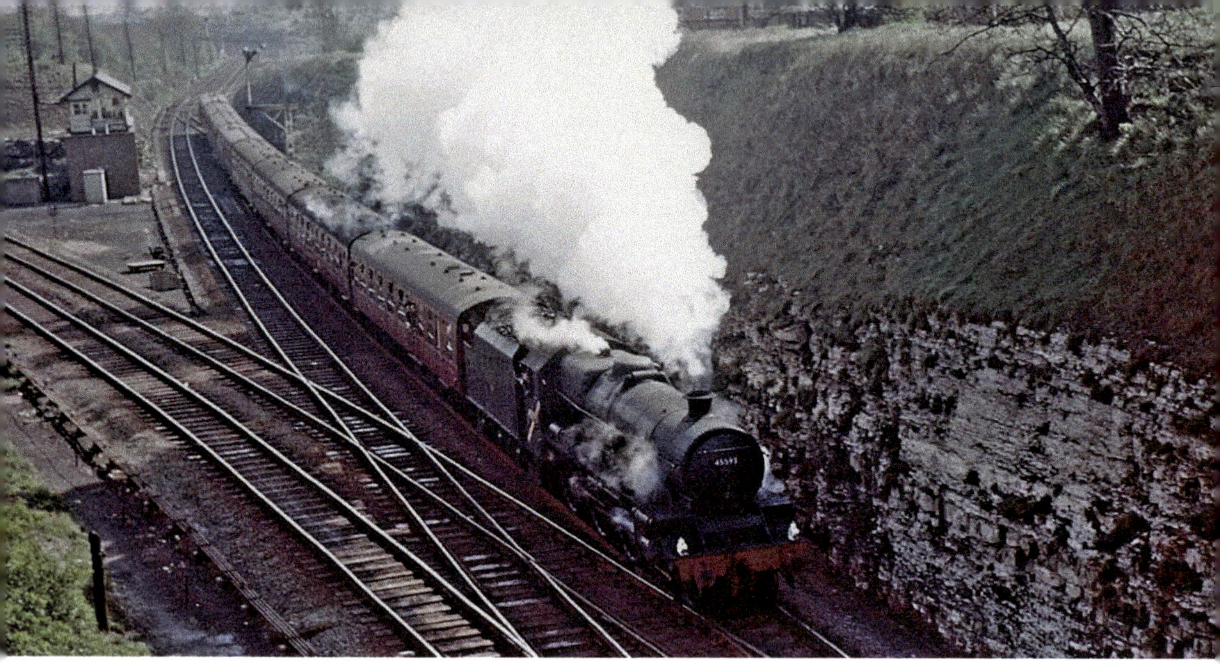

22 April 1967. LMS Jubilee 4-6-0 No. 45593 heads past Hexthorpe Junction.

LMS Jubilee 4-6-0 No. 45593 *Kolhapur* makes its way through Hexthorpe Junction with a special named the Derbyshire Dawdler. The train had originated from Leeds and was heading for Derby after an engine change at Chinley. The train was on the Doncaster to Sheffield line and the lines off to the left were the Doncaster avoiding lines, which diverted much of the heavy goods traffic away from Doncaster station and onto the lines to Thorne and beyond at Bentley Junction. (L. Flint)

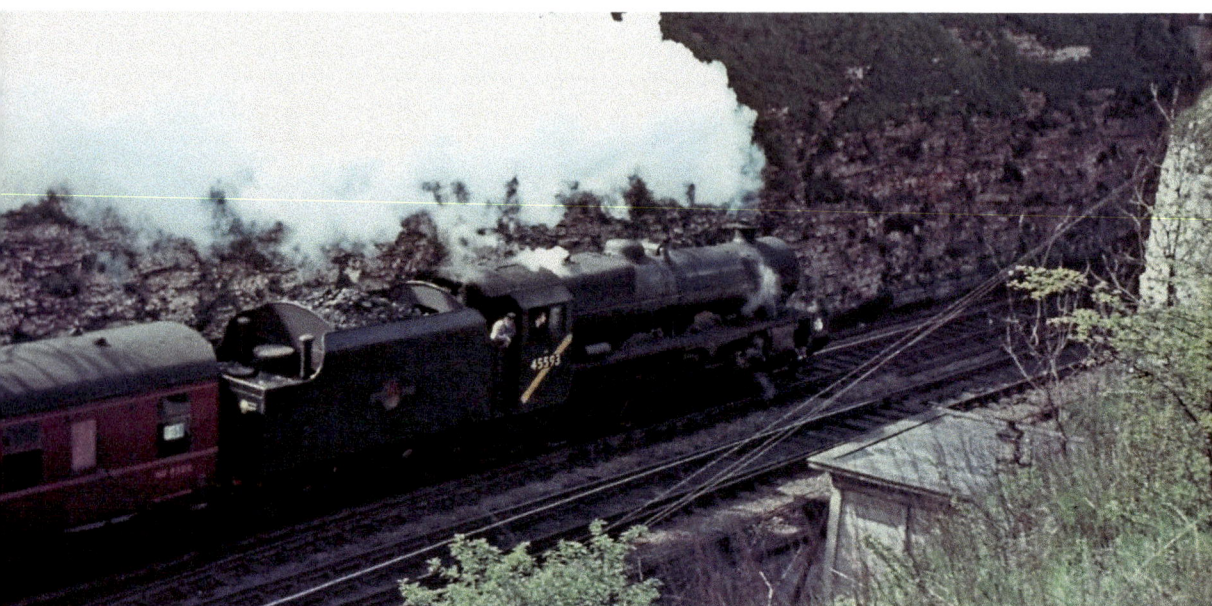

22 April 1967. LMS Jubilee 4-6-0 No. 45593 plunges into Warmsworth Cutting.

LMS Jubilee 4-6-0 No. 45593 *Kolhapur* plunges into Warmsworth Cutting with a special named the Derbyshire Dawdler. This was a deep-sided cutting with sheer limestone walls, which had originally been cut by the South Yorkshire Railway in 1847 and had since formed part of the line between Doncaster and Sheffield. The train was heading for Sheffield and then Chinley for a change of engines before arriving at Derby. (L. Flint)

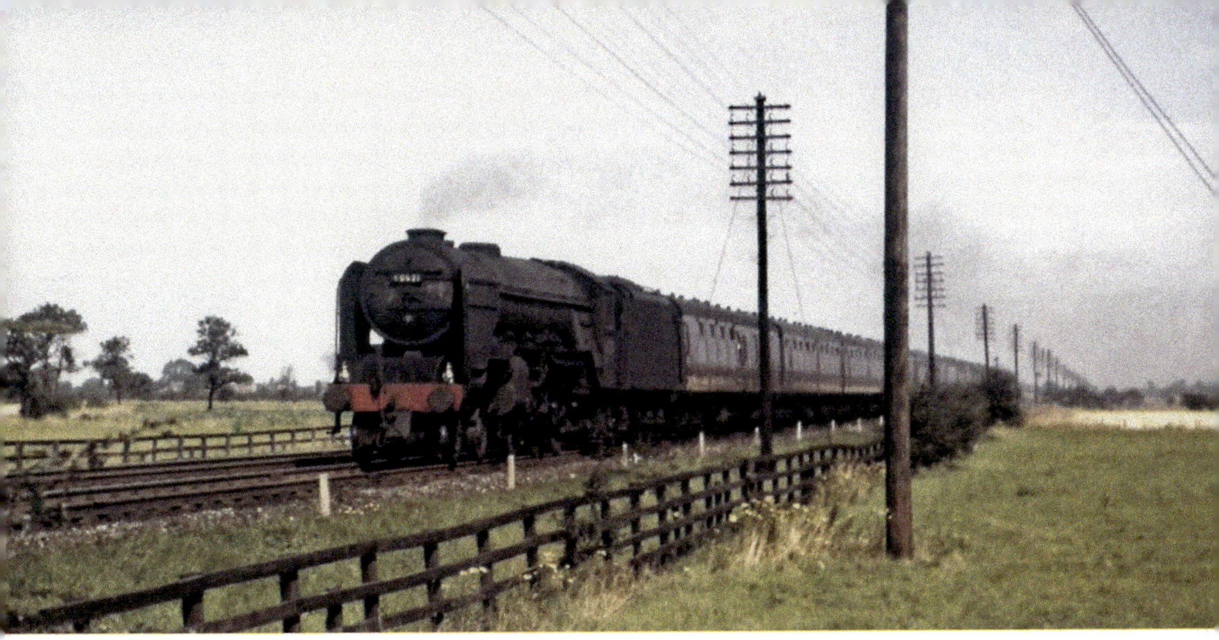

July 1962. LNER A2/2 4-6-2 No. 60521 races through Moss.

LNER A2/3 4-6-2 No. 60521 *Watling Street* was one of the fifteen pacifics built new to Edward Thompson's design. It had always been a North Eastern-allocated locomotive with time spent at both Gateshead and Heaton depots but spent its last year in service working from Tweedmouth MPD. It was seen passing south through the hamlet of Moss, north of Doncaster, on the East Coast Main Line. This location has very fond memories for me because in the late 1950s I would cycle, usually with a group of friends, the 7 miles from my home town to watch the spectacle of large express locomotives roaring along the flat countryside. (Photograph M. Fowler.)

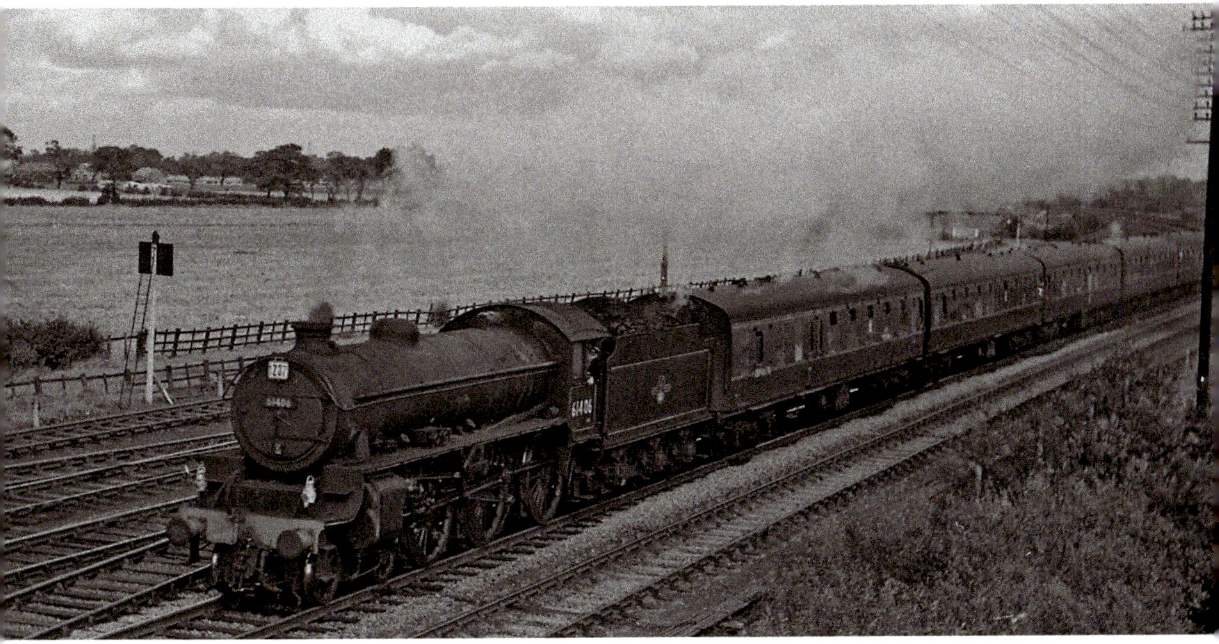

28 September 1963. LNER B1 4-6-0 No. 61406 heads through Thorne Junction.

LNER B1 4-6-0 No. 61406 was a very familiar locomotive to the local trainspotters in Thorne because it had been an Immingham-based engine from new until the depot closed in February 1966. It is seen hauling an excursion from the Cleethorpes and Thorne South direction past the large signal box at the junction heading towards Doncaster.

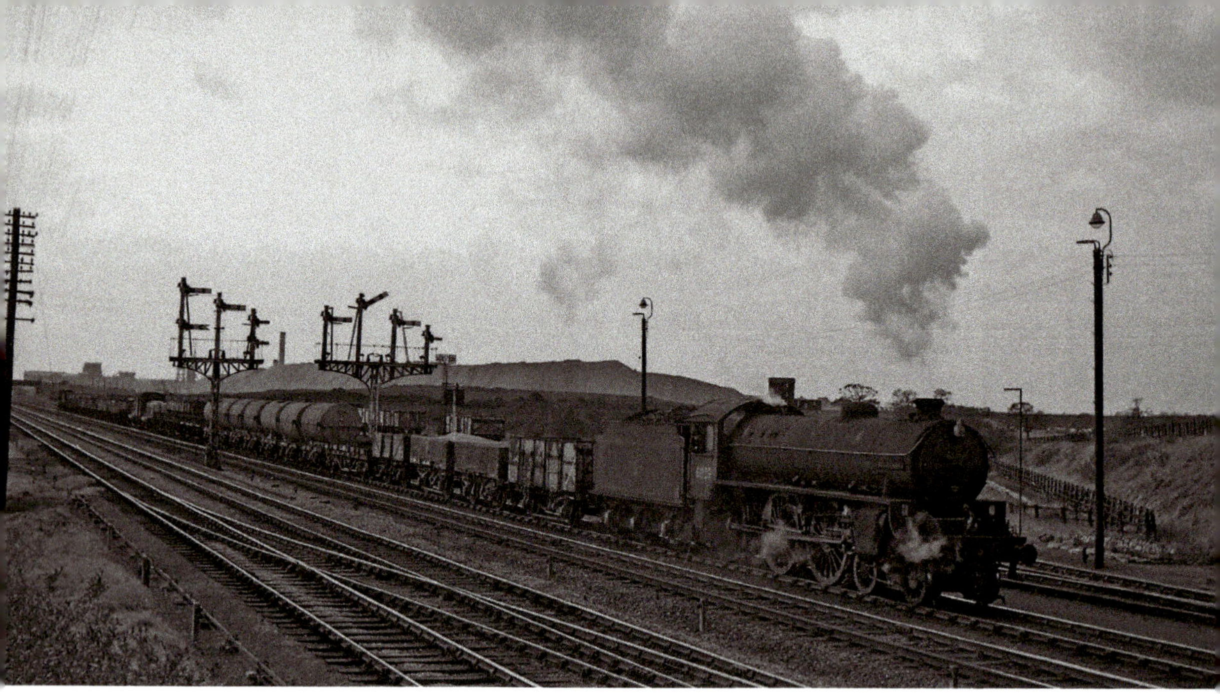

28 September 1963. LNER B1 4-6-0 No. 61189 approaches Thorne Junction.

LNER B1 4-6-0 No. 61189 *Sir William Gray* restarts a very mixed goods train towards Thorne Junction before taking the line towards Scunthorpe and Grimsby. It was a Mirfield-based locomotive at the time and its train would have most probably originated from the West Riding. The train had traversed the West Riding and Grimsby Railway link line from Adwick to Stainforth, which enabled trains to run directly from the Wakefield and Leeds area.

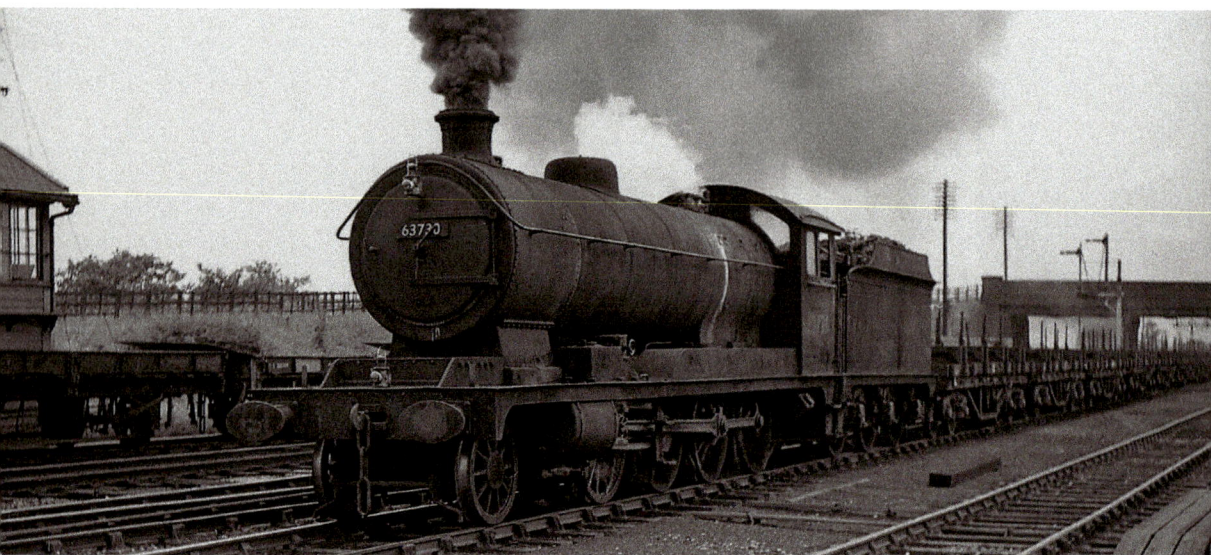

15 August 1964. LNER O4/8 2-8-0 No. 63730 restarts from Thorne Junction.

LNER O4/8 2-8-0 No. 63730 restarts a long string of empty bogie bolster wagons through Thorne Junction. It would take the slow line towards Thorne South station and then on to Scunthorpe. It is seen passing a train of steel plate heading in the opposite direction. This was typical of the type and volume of freight traffic at the time. The engine had been built for the Railway Operating Division of the Royal Engineers in November 1919 and eventually bought by the LNER in April 1924. It was rebuilt in March 1958 as an O4/8 with a B1 type boiler and cab and spent its last three years working from Doncaster MPD. It was withdrawn in January 1966 and cut up by T. W. Ward at Beighton.

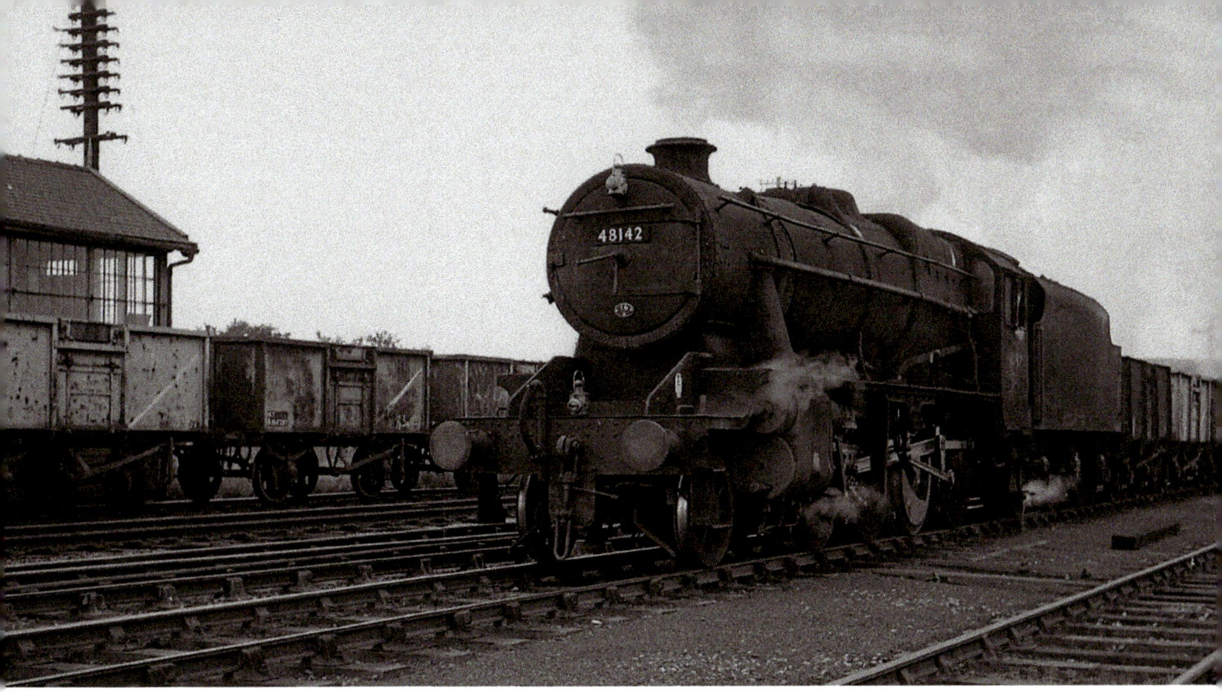

15 August 1964. LMS 8F 2-8-0 No. 48142 heads through Thorne Junction.

LMS 8F 2-8-0 No. 48142 hauls a loaded coal train passed Thorne Junction signal box as a train of empty mineral wagons heads in the opposite direction. The 8F was allocated to Annesley MPD at this time and was hauling coal from the Nottinghamshire coalfields towards Scunthorpe. These locomotives were regular performers along the old GCR route through Thorne South but were more usually seen hauling iron ore trains to Scunthorpe.

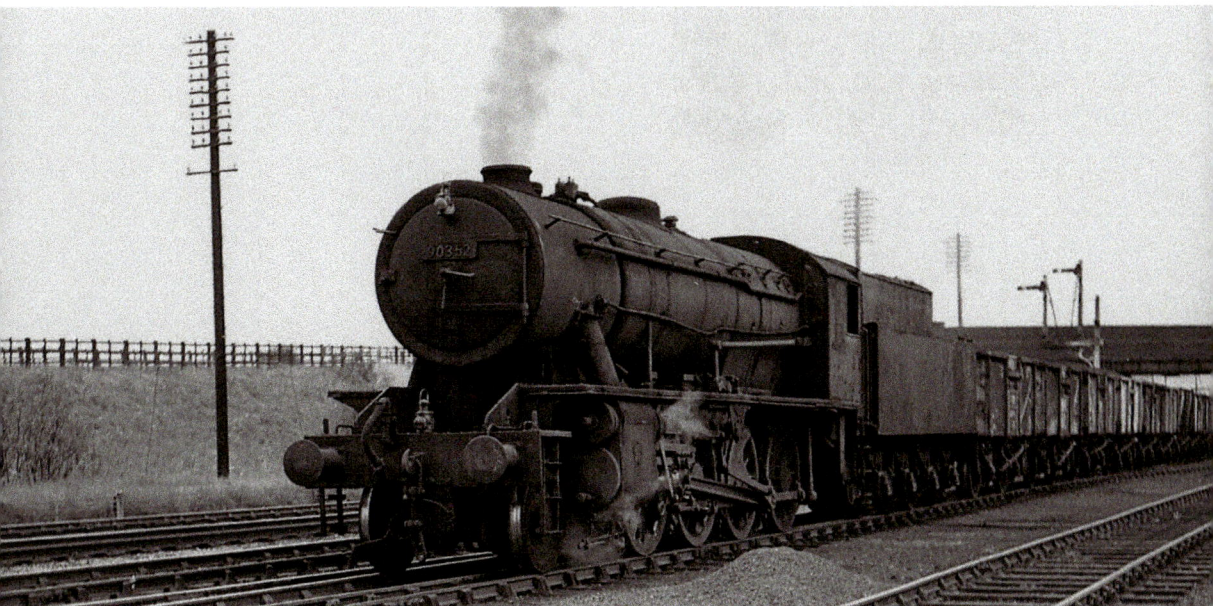

15 August 1964. WD 2-8-0 No. 90352 hauls a coal train through Thorne Junction.

WD 2-8-0 No. 90352 heads a coal train back to its home city of Hull from the South Yorkshire coalfields. Trains of coal were constantly heading to the ports of Goole and Hull for export to many far-flung countries. This locomotive had spent the last fourteen years doing this type of work and had two more years of service before it was withdrawn and scrapped in Hull at the yard of Albert Draper.

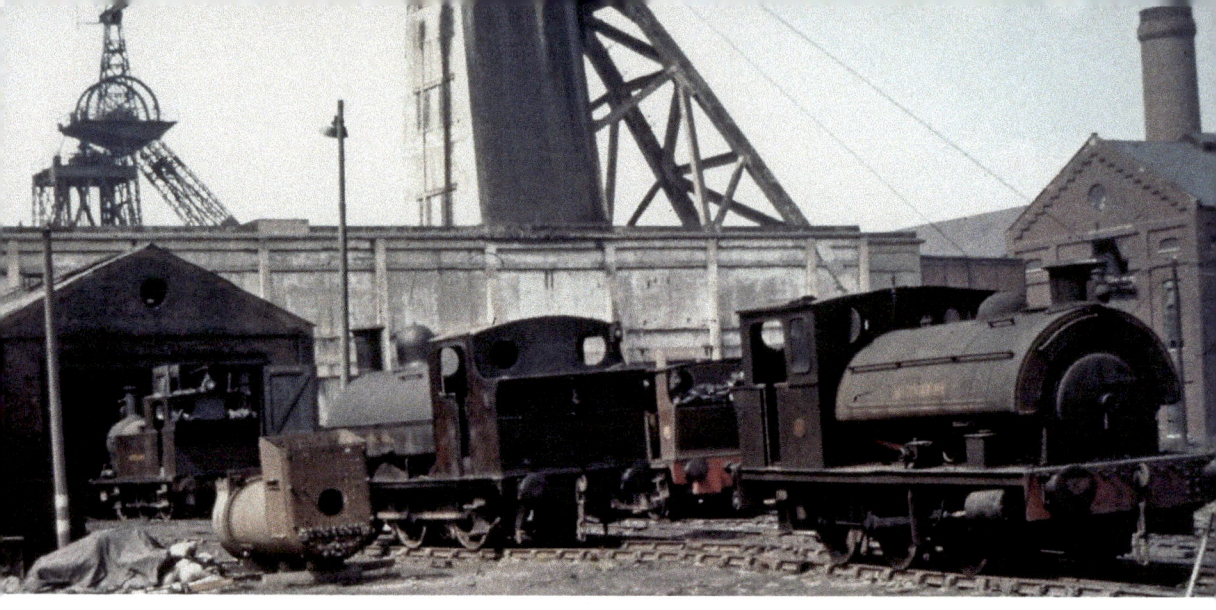

April 1961. Hatfield Colliery locomotive shed and yard.

The variety of motive power on display in the shed yard at Hatfield Colliery was diverse and of great interest. The front engine on the shed road was 0-6-0ST *Hatfield No. 1*, a Hawthorn Leslie product of 1916 (works number 3197). Next in line was 0-6-0ST *Hatfield No. 3*, a Kitson-built locomotive (works number 3819) of 1899 for the Taff Vale Railway, which became GWR No. 791 in 1922. *Hatfield No. 4* in front of the shed was built in 1902 by Avonside for the Llanelly & Mynydd Mawr Railway. Finally the locomotive behind, with just the bunker visible, was 0-6-0ST *Hatfield No. 6*, a Hudswell Clarke-built engine (works number 1349) of 1918 vintage. (L. Flint)

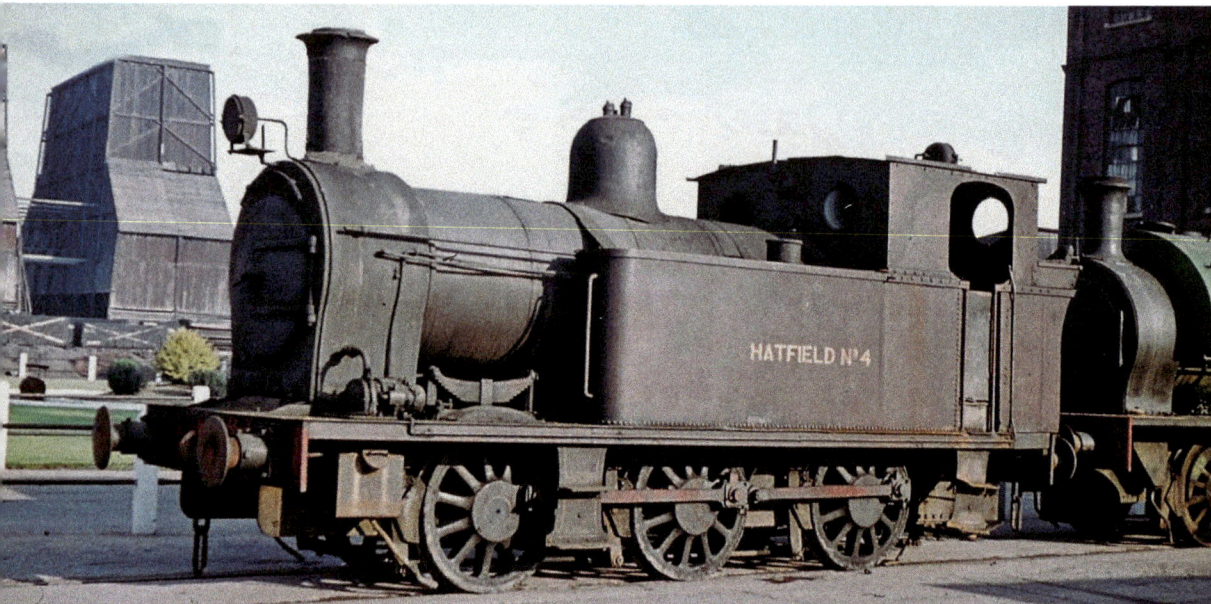

June 1961. Avonside 0-6-0T *Hatfield No. 4* at Hatfield Colliery.

Avonside 0-6-0T *Hatfield No. 4* was built in 1902 by Avonside (works number AE 1448) for the Llanelly & Mynydd Mawr Railway, which was taken over by the GWR and the engine number was changed to No. 944. It was withdrawn in November 1928 by the GWR and sold early in 1929 for further use to Carlton Collieries Associates. It was allocated to Hatfield Main Colliery Co. Ltd, becoming *Hatfield No. 4*. After forty years service at Hatfield it was sold for scrap by the National Coal Board in early 1968.

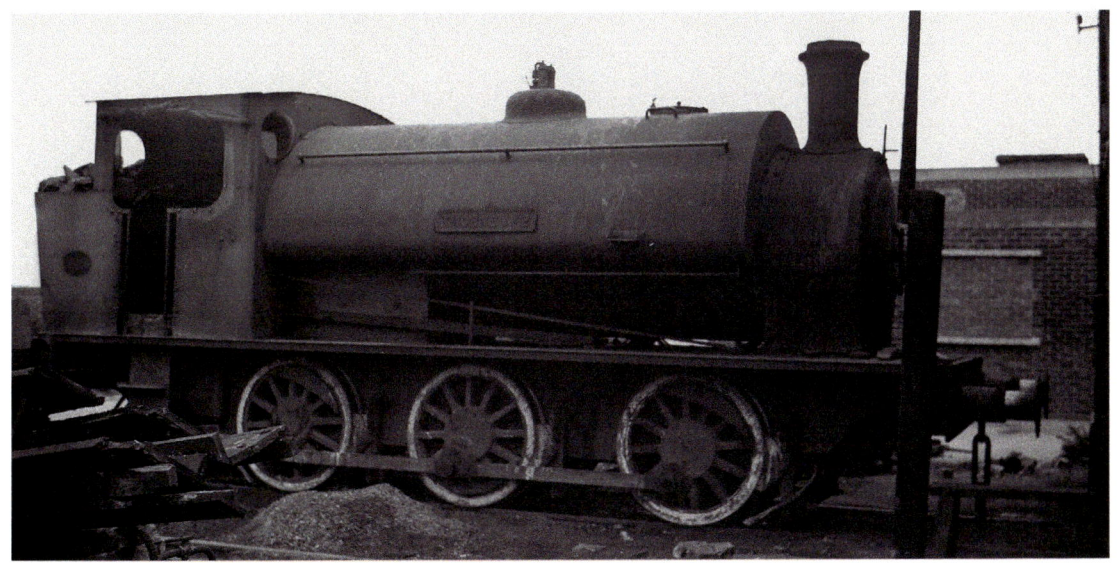

October 1962. Hunslet 0-6-0ST No. 3804 *Thorne No. 2* at Thorne Colliery.

Hunslet 0-6-0ST (works number 3804) *Thorne No. 2* was built in 1953 and was one of two similar locomotives based at NCB Thorne Colliery. Although the colliery had ceased producing coal in 1956 because of flooding problems, some activities continued on the surface requiring the use of shunting locomotives. *Thorne No. 2* was to be transferred to the nearby Hatfield Colliery by 1964 and continued its work there until replaced by a redundant British Railways diesel shunting locomotive.

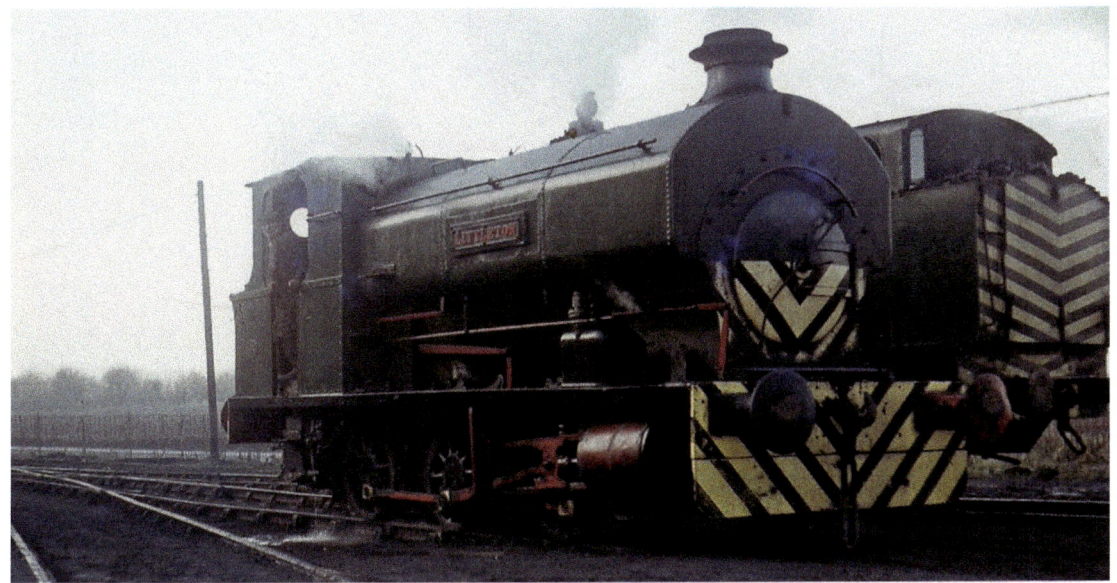

March 1963. 0-6-0ST Avonside *Littleton* at Askern Main Colliery.

0-6-0ST *Littleton* was an Avonside engine (works number 1833) that had been built in 1919. It is seen in the shed yard at Askern Colliery, which operated a variety of 0-6-0 tank locomotives. Whether the name on this engine referred to Littleton Colliery in Cannock, Staffordshire, or not I have not been able to establish. That colliery had a range of locomotives that had similar nameplates but also a number plate below it. They all seemed to have been Manning Wardle products though.

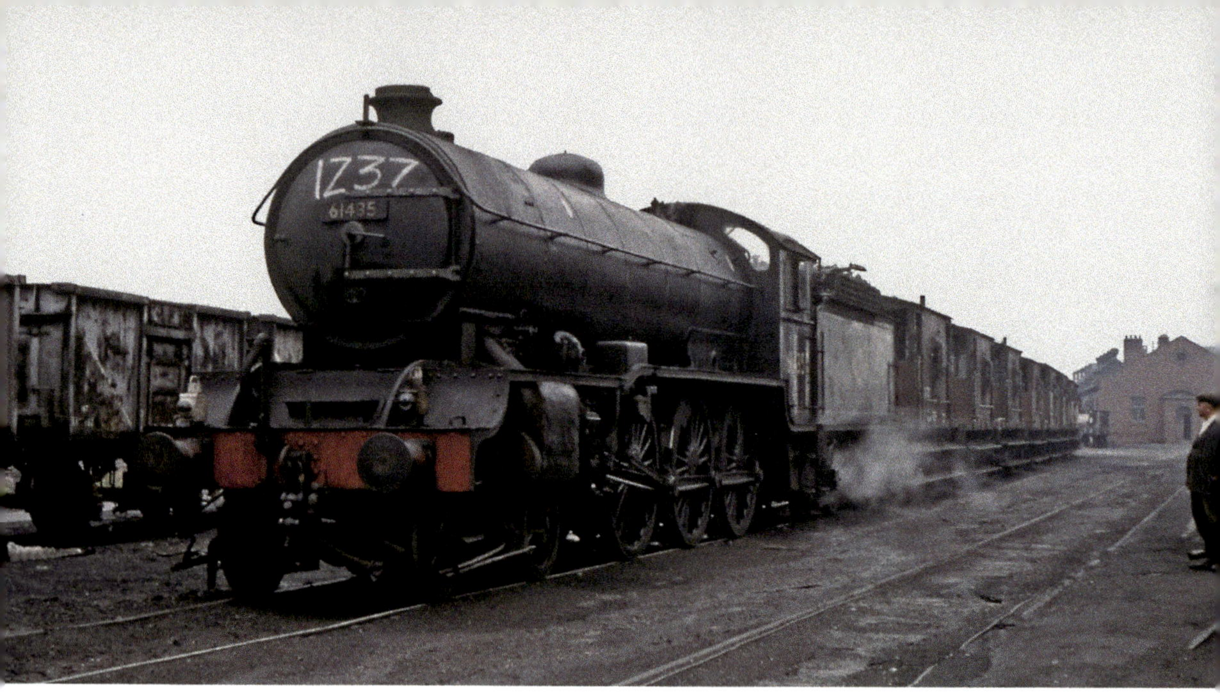

18 July 1964. LNER B16 4-6-0 No. 61435 at Harworth Colliery.

LNER B16/2 4-6-0 No. 61435 was used on a brake van enthusiasts' special that had been organised by the Gresley Society for its members. The special had visited many lines in the Doncaster area and then spent time at Harworth Colliery to see the society's newly acquired N2 0-6-2T No. 69523. It was stored at the colliery having been rescued from the scrap line in Doncaster Plant Works.

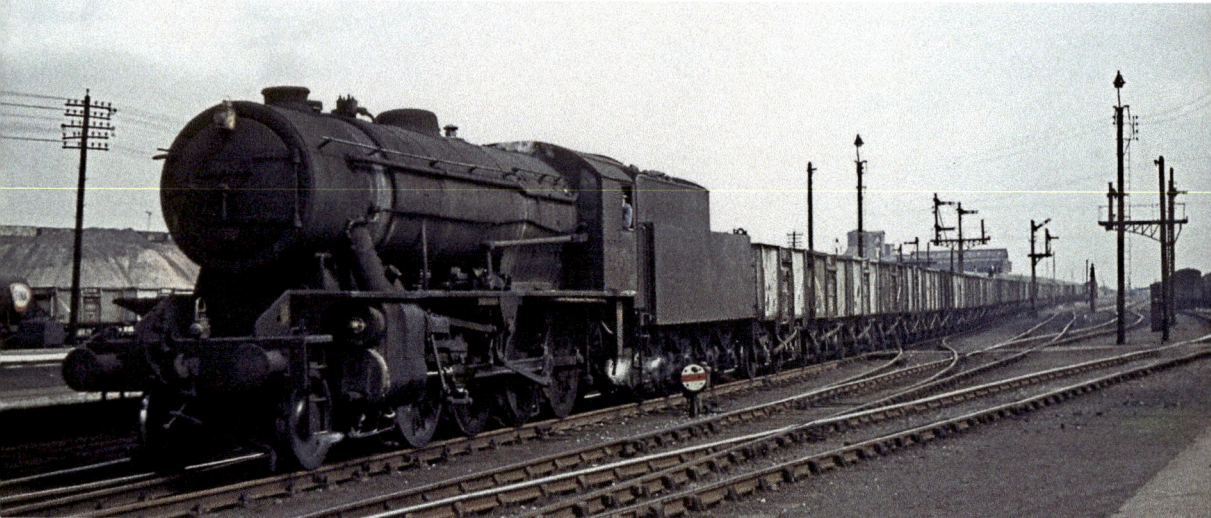

15 May 1964. WD 2-8-0 No. 90456 at Stainforth.

WD 2-8-0 No. 90456 was built April 1944 by Vulcan Foundry as WD No. 78642. It was one of a number of 2-8-0s allocated to the United States Transportation Corps for use in north-west Europe but by November 1945 had been loaned to the LNER as their No. 3135, which was allocated to New England MPD. It was one of 200 2-8-0s purchased by the LNER for £4,500 each in November 1946 and by June 1948 had been reallocated to Immingham. Its final move came in May 1951 when it became a resident of Frodingham MPD and remained there until withdrawal in February 1966 after which it was scrapped at Draper's of Hull. It was seen heading south through Stainforth with yet another string of coal empties. The slag heap and buildings of Hatfield Main Colliery are visible in the background.

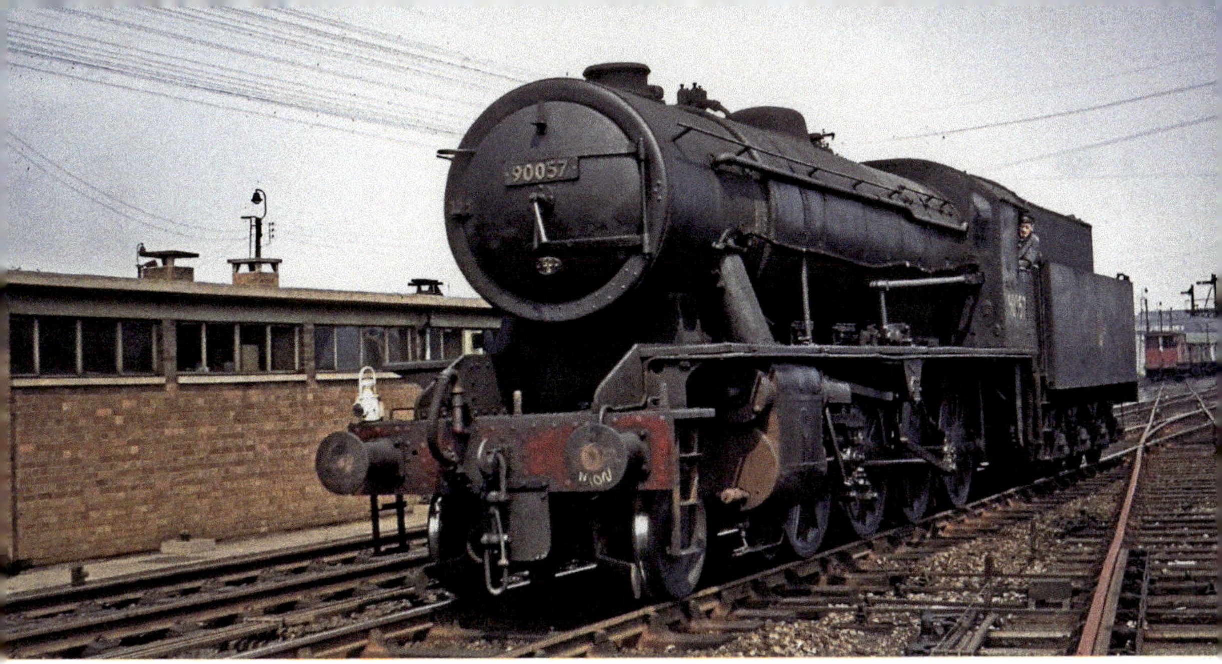

4 June 1965. WD 2-8-0 No. 90057 at Stainforth.

WD 2-8-0 No. 90057 was built May 1944 by the North British Locomotive Company of Glasgow as WD 70828. By November 1945 it was loaned to the LNER as No. 3057 working from Colwick MPD and was one of 200 of the class purchased by the LNER in November 1946. Apart from a year at Tyne Dock in 1963, No. 90057 spent most of its British Railways career working out of Hull. Its last six months of service were at Goole MPD after which it was withdrawn in June 1967. It was seen heading light engine through Stainforth.

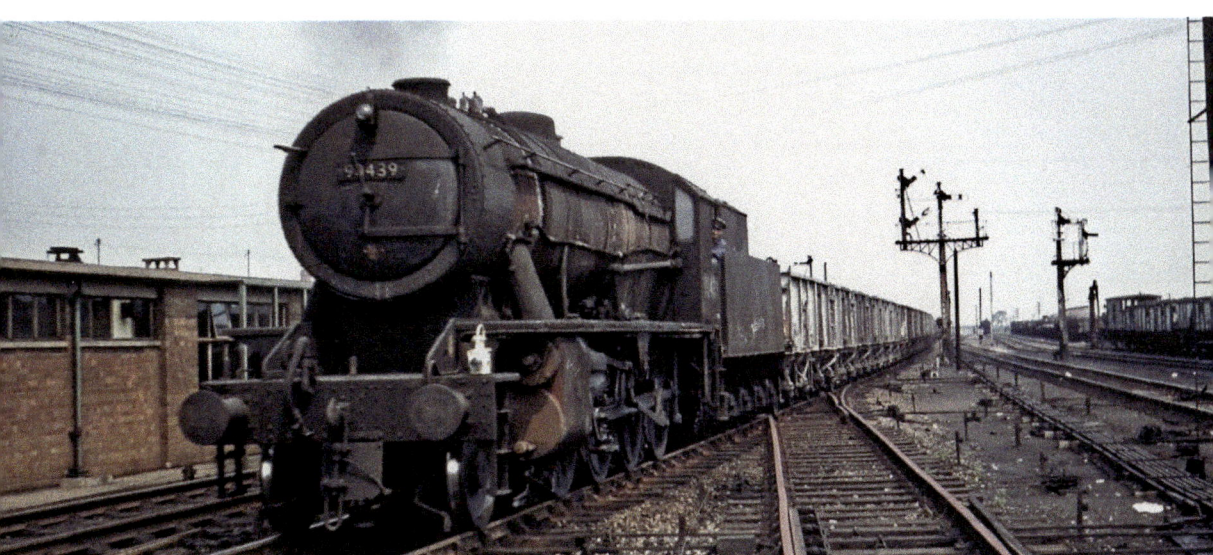

4 June 1965. WD 2-8-0 No. 90439 at Stainforth.

WD 2-8-0 No. 90439 was built March 1944 by Vulcan Foundry as WD No. 77496. After being shipped to Belgium it was put to work on the SNCF, the French State railway, but was back in the UK and on loan to the LNER at New England MPD by May 1946. In March 1947 it was purchased by the LNER and became No. 3118. With the formation of British Railways in 1948 it became No. 90439 and by April 1963 had joined the books of Frodingham MPD where it remained until withdrawn in December 1965. It is seen heading a train of mineral empties through Stainforth station.

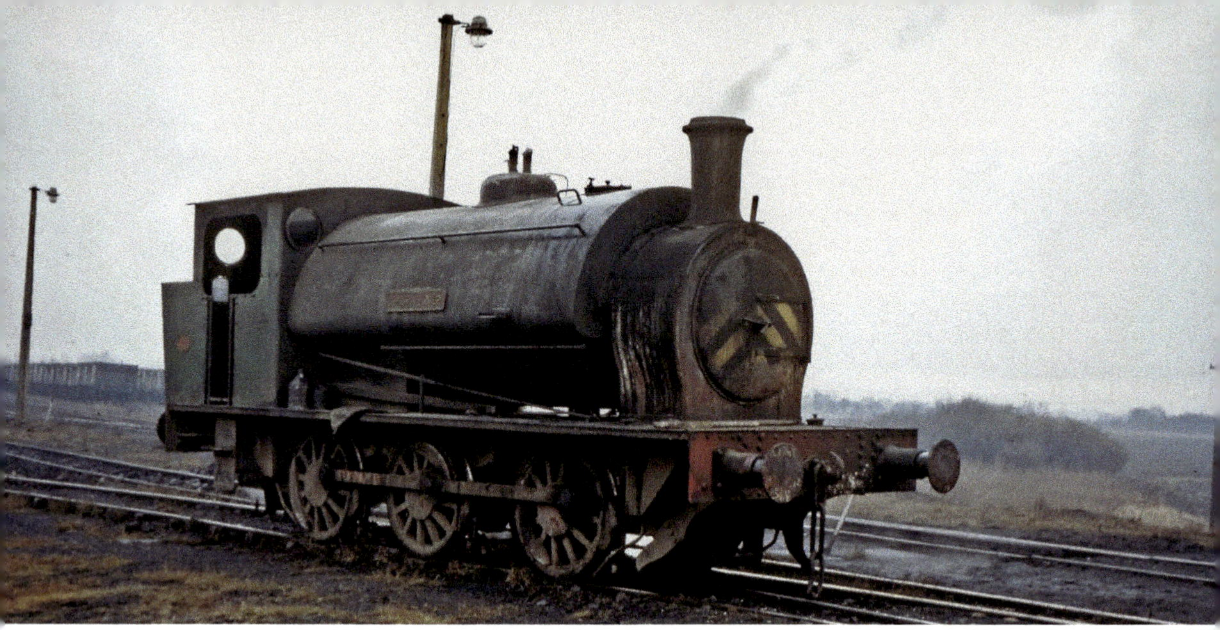

January 1967. 0-6-0ST *Thorne No. 2* at Thorne Junction.

Hunslet 0-6-0ST (works number 3804) *Thorne No. 2* was built in 1953. It had been transferred from the closed Thorne Colliery to the nearby Hatfield Colliery by 1964 and continued its work there until replaced by an ex-British Railways diesel shunter. The engine is standing at the most northerly end of the vast complex of colliery sidings, which included a branch to the Stainforth and Keadby Canal. Here the coal was transferred from railway wagons to barges and the famous Tom Puddings tubs, which transported it to the Port of Goole from where it was shipped by sea.

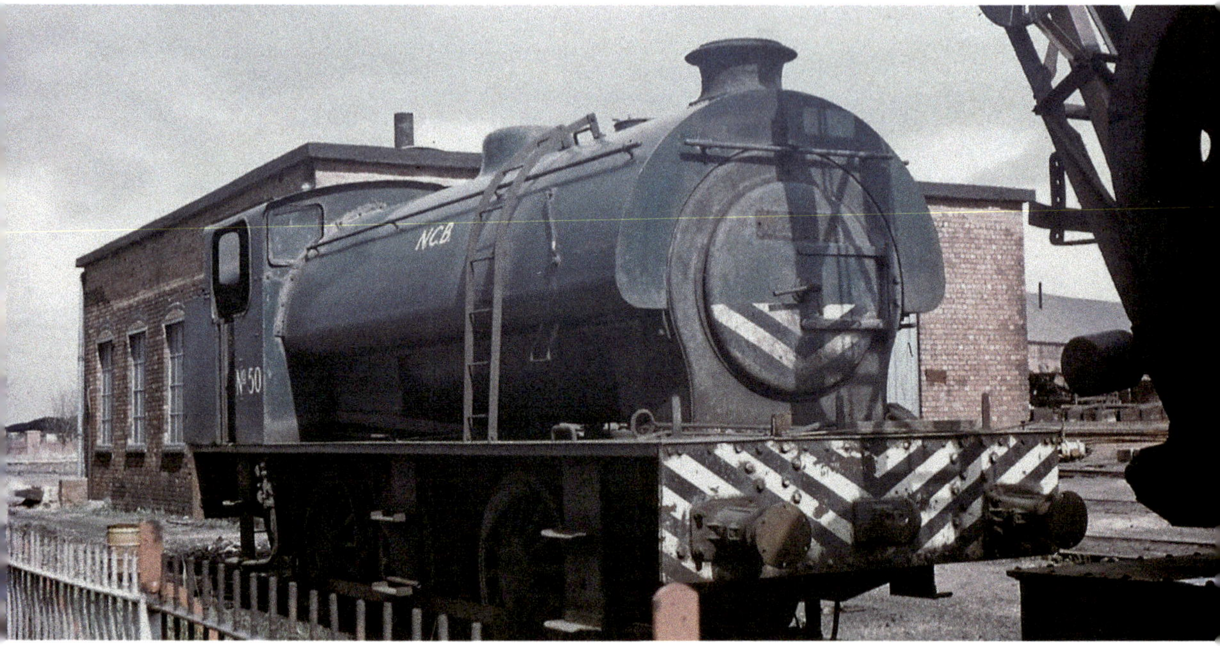

18 April 1970. LNER J94 0-6-0ST No. 68020 at Askern Main Colliery.

LNER J94 0-6-0ST No. 68020 had been built by Bagnall (works number 2752) in 1944 for the War Department and was one of seventy-five purchased by the LNER in 1946. It had been withdrawn from Langwith Junction MPD in June 1963 and was overhauled at the Plant Works in Doncaster before entering service with the National Coal Board. It became NCB No. 50 at Askern Main Colliery and was cut up in May 1970.

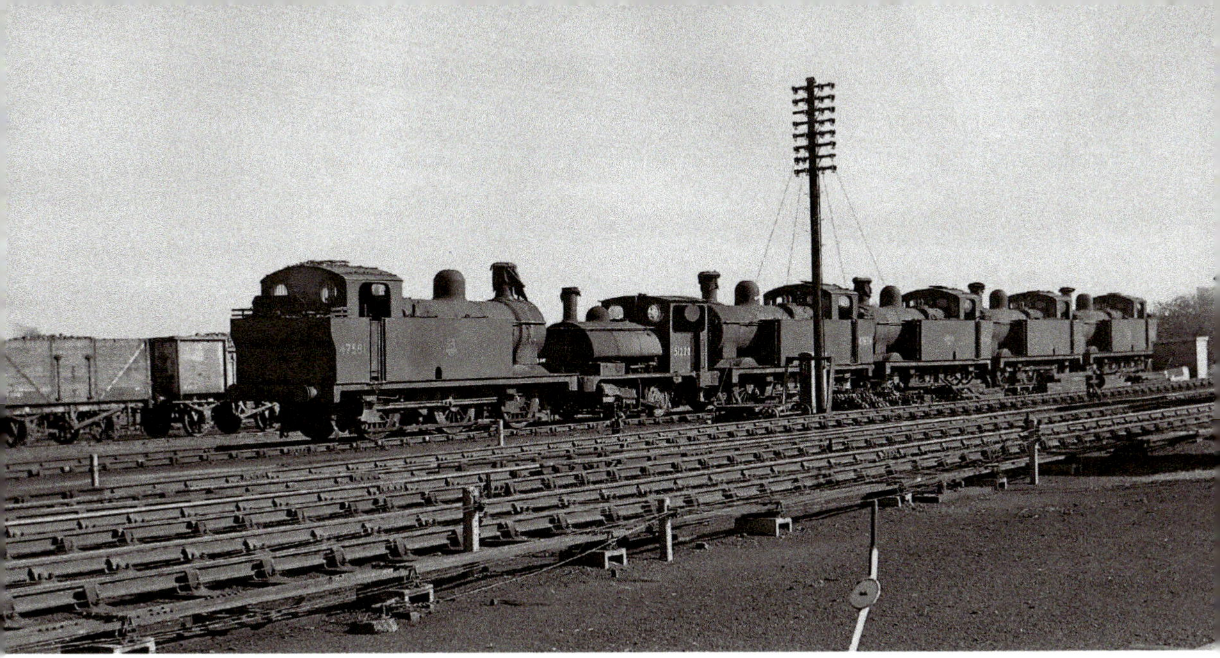

April 1961. A line of stored tank locomotives at Goole MPD.

LMS 0-6-0T 3F Jinties Nos 47581, 47634, 47438, 47589 and 47462 were stored at Goole MPD along with L&YR Class 21 0-4-0ST No. 51222 in the spring of 1961. The Jinties (except No. 47589) had moved to Goole via York after Harrogate's Starbeck shed had closed but they found no work at the port. Two of the group, Nos 47581 and 47589, were eventually reallocated to Farnley Junction, while the others were to linger on in store until the summer of 1962 when they were removed for scrap.

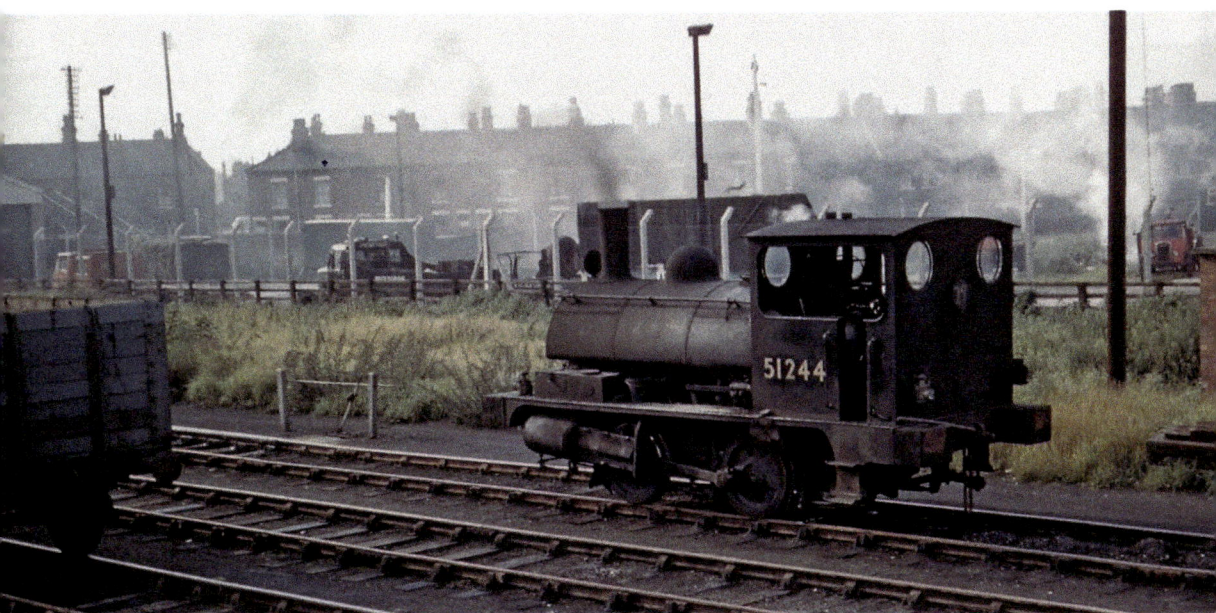

August 1960. L&YR Class 21 0-4-0ST No. 51244 shunting in Goole.

Lancashire & Yorkshire Railway Class 21 0-4-0ST No. 51244 rests between shunting duties on the dock sidings in the Stanhope Street area of Goole. Behind the locomotive is the British Road Services depot on Mariners Street, a scene which has been radically altered over the intervening years. The engine would survive at Goole, with several periods in store, before being loaded onto a bogie well wagon and scrapped at Crewe in August 1962.

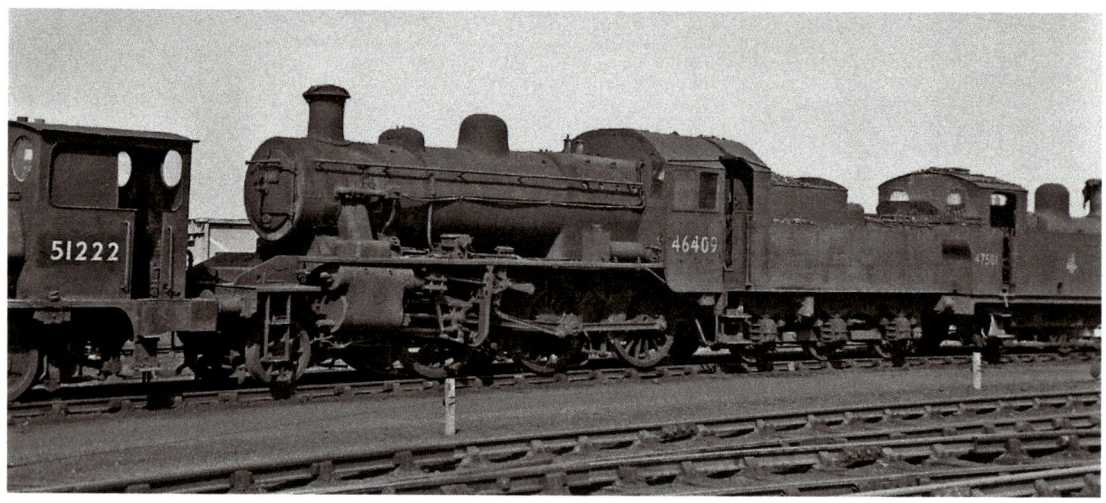

28 August 1961. Ivatt 2MT 2-6-0 No. 46409 stored at Goole MPD.

LMS 2-6-0 Ivatt 2MT No. 46409 was one of a number of the class allocated to Goole in the early 1960s. They were used on pickup goods trains to Thorne, Pontefract, Selby and on the Isle of Axholme, as well as local train services to Wakefield. By 1961 much of their work had been lost to diesel railcars and shunters. This locomotive eventually found work elsewhere though, in January 1962 at Scarborough before returning to Goole for a year and finally being transferred to Hurlford from where it was withdrawn in 1964. (Photograph J. B. Platt)

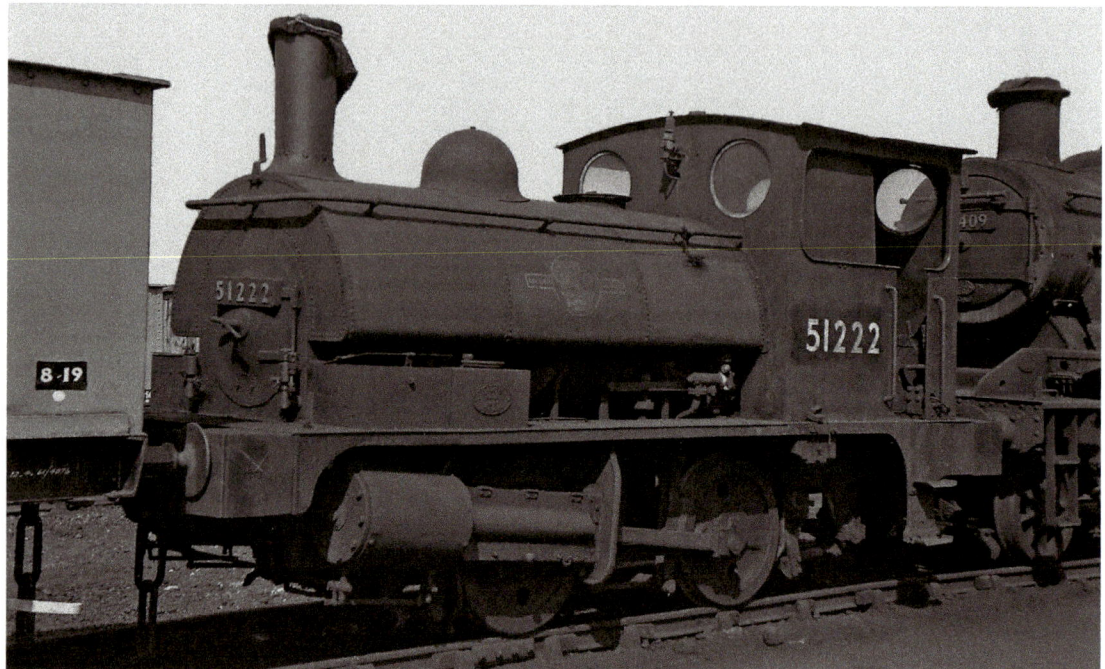

28 August 1961. L&YR Class 21 0-4-0ST No. 51222 stored at Goole MPD.

Lancashire and Yorkshire Railway Class 21 0-4-0ST No. 51222 was one of three allocated to Goole in the late 1950s and early 1960s and were used in the docks and along Aire Street in Goole. Although officially withdrawn in March 1962, it was already in store in the summer of the previous year, because diesel replacements were being used to do the work. (Photograph J. B. Platt)

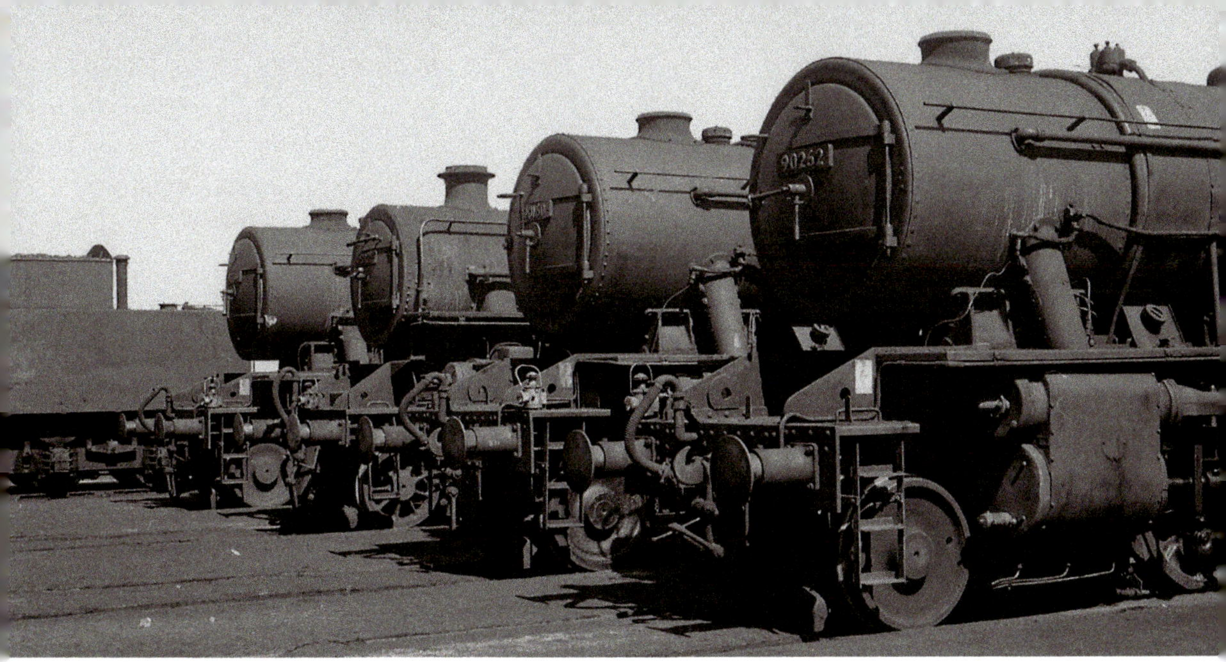

April 1962. Locomotives line up at Goole MPD.

WD 2-8-0s dominated the scene at Goole with four in this photograph. Of the two WDs identified, the nearest, No. 90262, was a Goole loco for thirteen years but would end its working life at Hull. Next in line was No. 90160, another Goole engine for ten years before finally being withdrawn from Hull also. The third locomotive in the line was LMS Ivatt 2-6-0 4MT No. 43125, which was a Goole resident for six years before finding a couple of years of work in West Yorkshire.

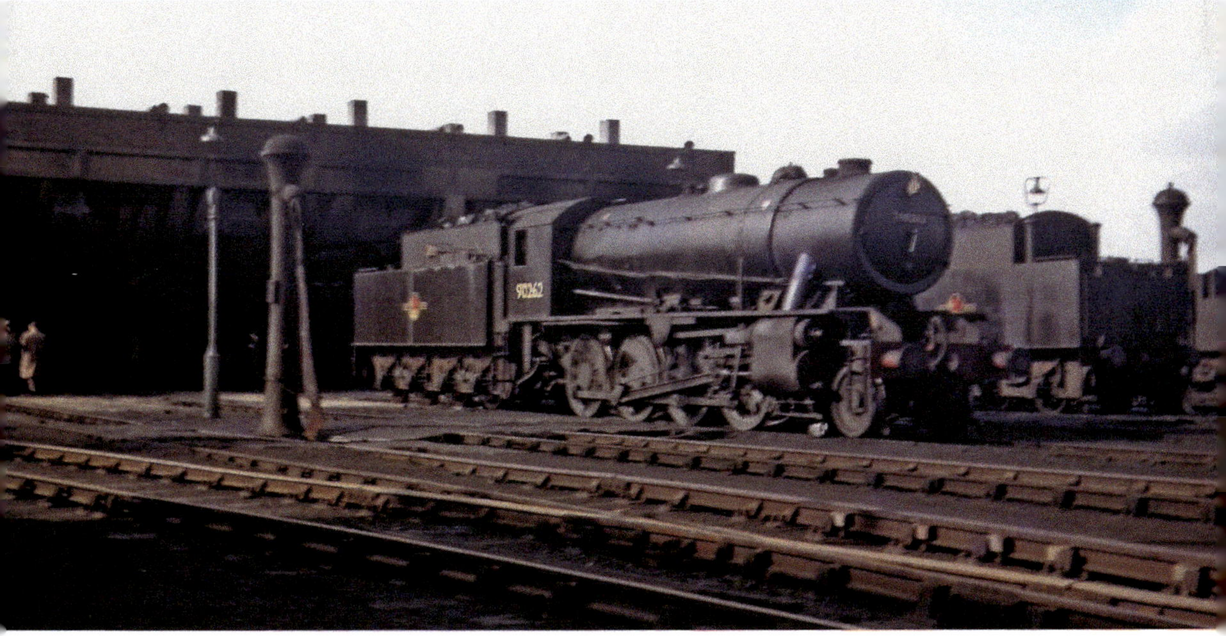

February 1964. WD 2-8-0 No. 90262 on Goole MPD.

WD 2-8-0 No. 90262 had been a Goole-based locomotive since 1950 but had moved to Hull Dairycoates MPD a few months before this photograph was taken. It had recently returned to service after having a general overhaul at the Plant in Doncaster. It was to continue working from Hull until withdrawal in June 1967.

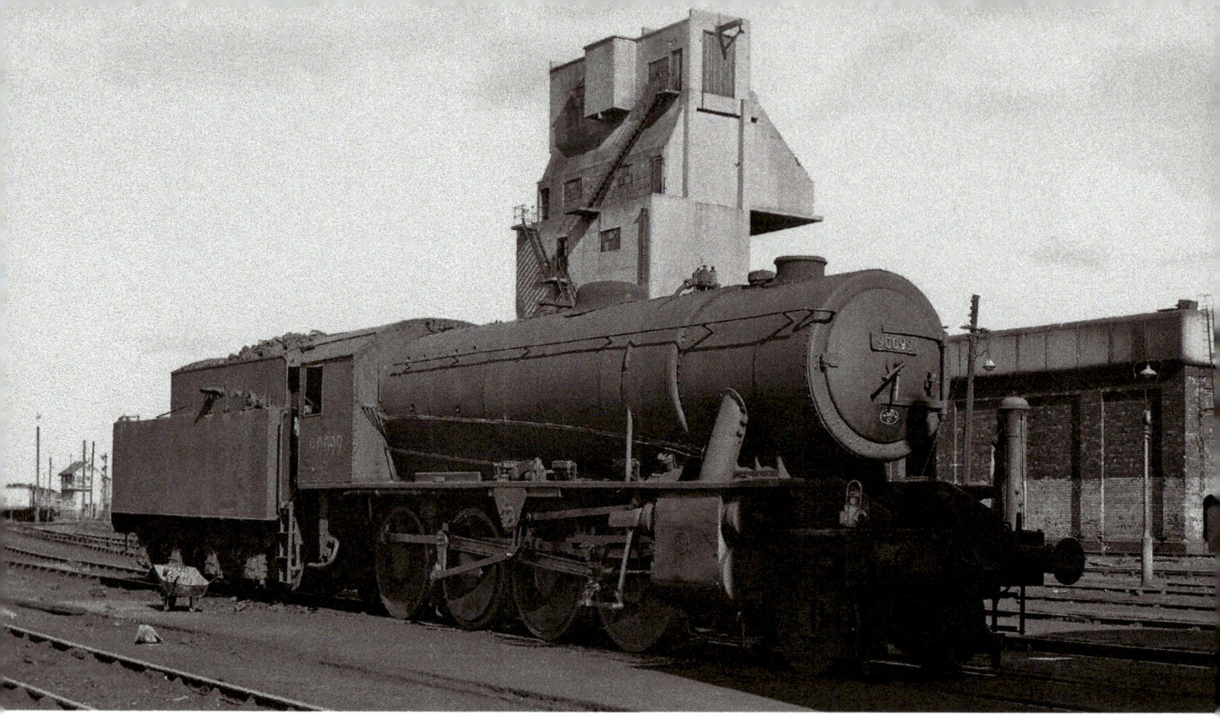

May 1966. WD 2-8-0 No. 90099 on Goole MPD.

WD 2-8-0 No. 90099 rests on shed at Goole in 1966 by which time this class was predominant at the shed. It had been a local engine all its working life, coming to Goole from Hull Dairycoates in 1963. It remained a Goole engine until finally being withdrawn in June 1967 and broken up at Draper's in Hull.

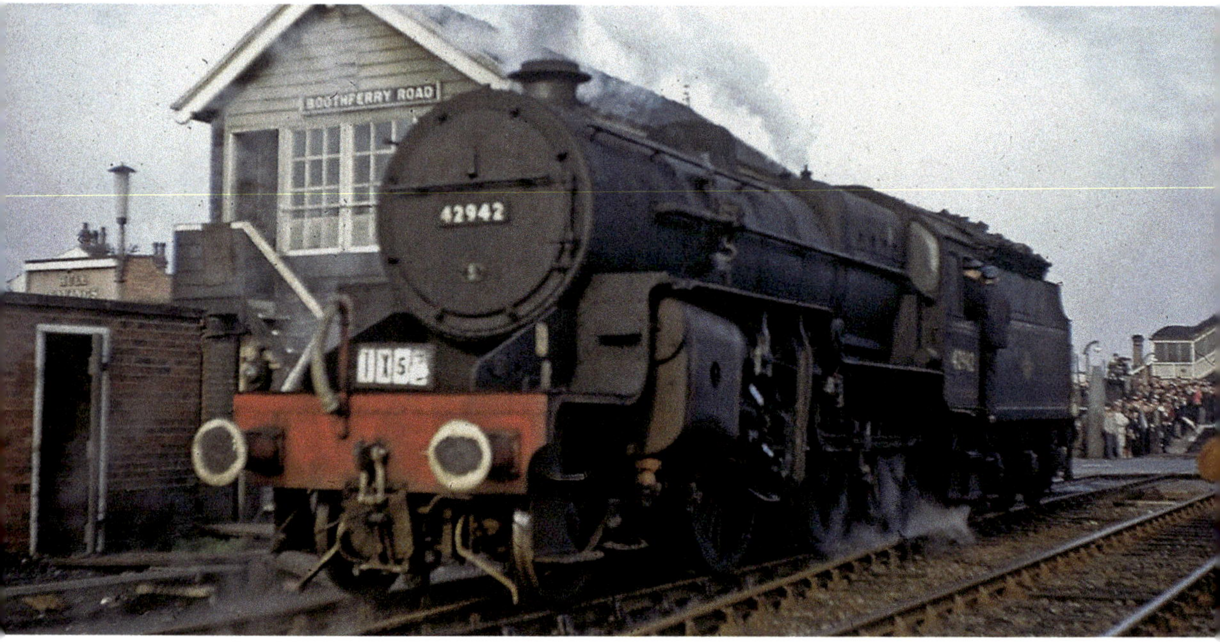

8 October 1966. LMS Crab 2-6-0 No. 42942 at Goole.

LMS Crab 2-6-0 No. 42942 backs onto the LCGB Crab Commemorative Railtour at Goole station. While the WD had brought the special from Wakefield Kirkgate, the Crab had been watered, coaled and turned at Wakefield before following on light engine over an hour later.

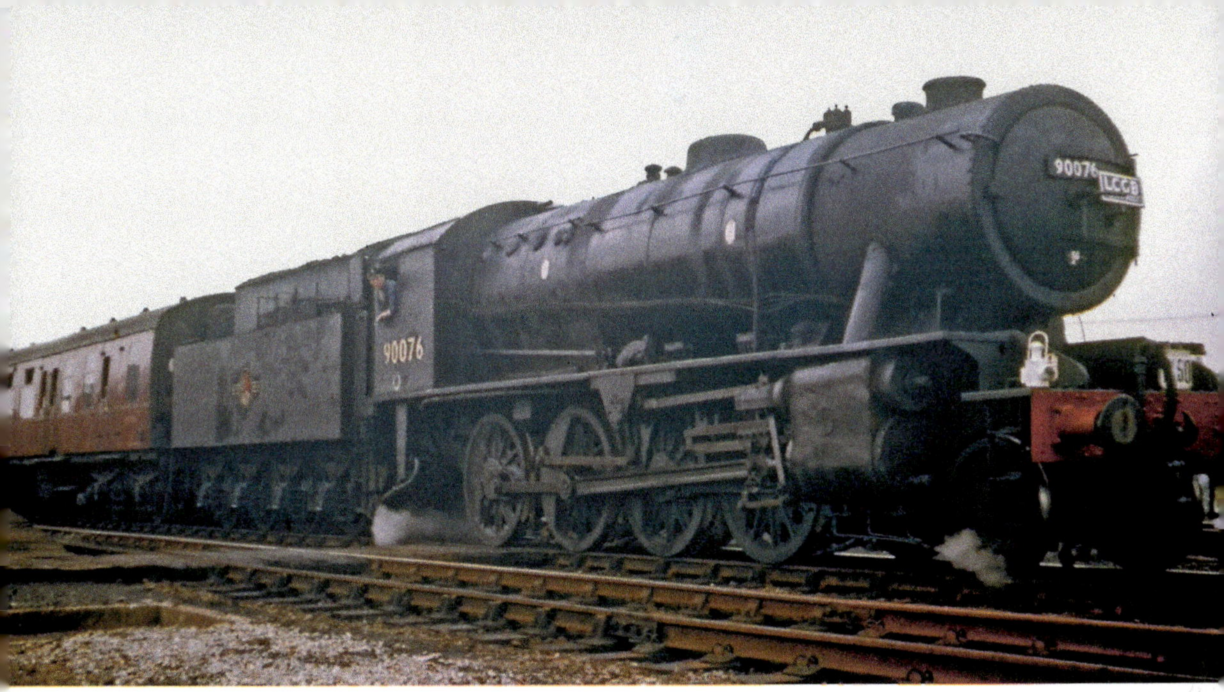

8 October 1966. WD 2-8-0 No. 90076 at Goole.

WD 2-8-0 No. 90076 arrived at Goole station with the LCGB Crab Commemorative Railtour, which it had brought from Wakefield Kirkgate. It had been specially cleaned for the occasion and covered the journey in a little over one hour. The WD had been a Wakefield-allocated engine since 1959 but spent the last month of its active life at West Hartlepool MPD and was withdrawn in September 1967.

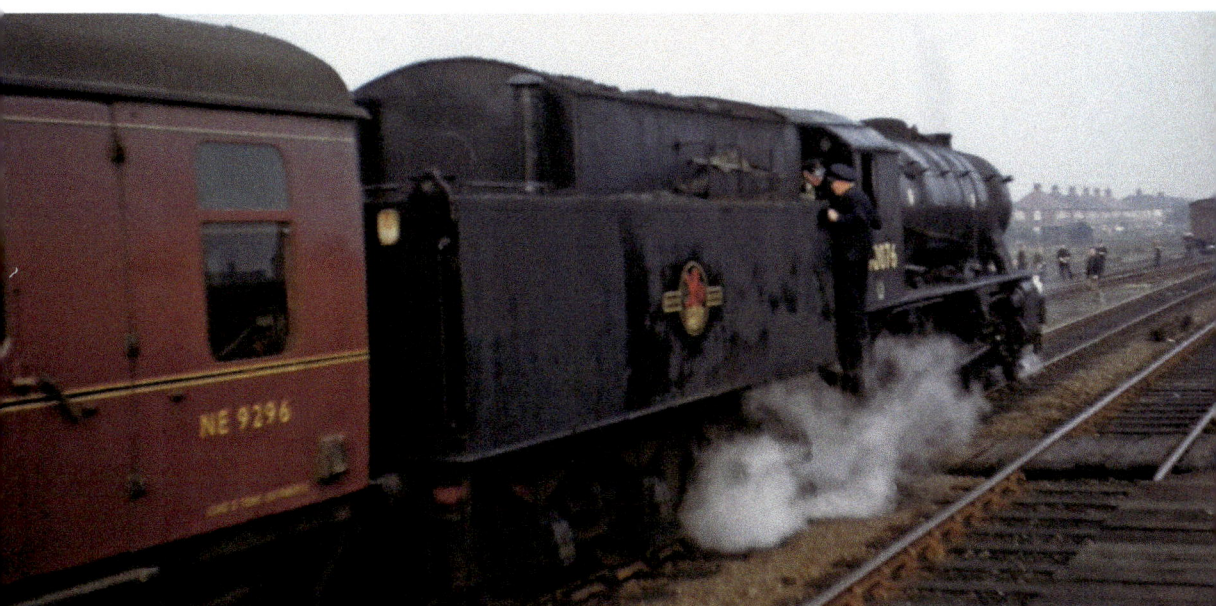

8 October 1966. WD 2-8-0 No. 90076 arrives at Goole.

WD 2-8-0 No. 90076 attracted a lot of attention at Goole station having arrived with the LCGB Crab Commemorative Railtour, which it had brought from Wakefield Kirkgate. The train had originated from Liverpool Exchange and had included much of the Lancashire and Yorkshire network on its journey. The WD would be replaced by LMS Crab 2-6-0 No. 42942 for the return journey to Liverpool.

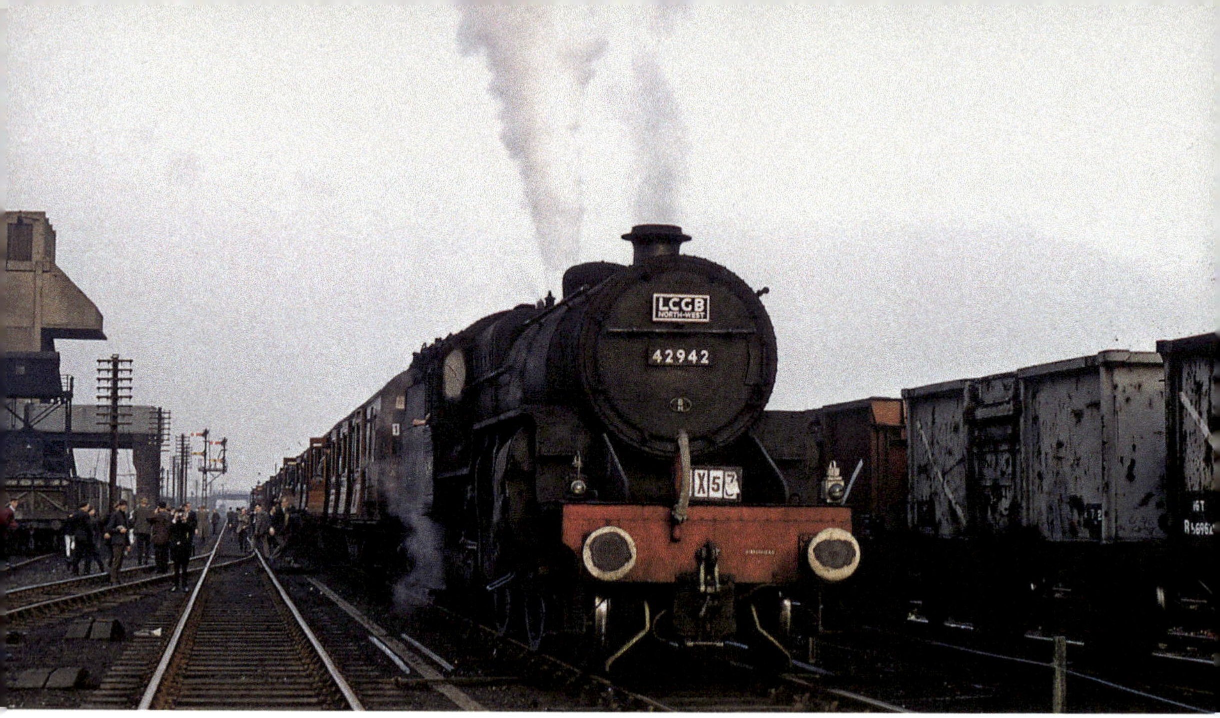

8 October 1966. LMS Crab 2-6-0 No. 42942 at Goole.

LMS Crab 2-6-0 No. 42942 halts its LCGB Crab Commemorative Railtour opposite Goole MPD. The special left Goole to Knottingley line using the route via Potters Grange Junction and Engine Shed Junction. A hastily arranged unofficial shed visit was made with passengers detraining and walking across the running lines.

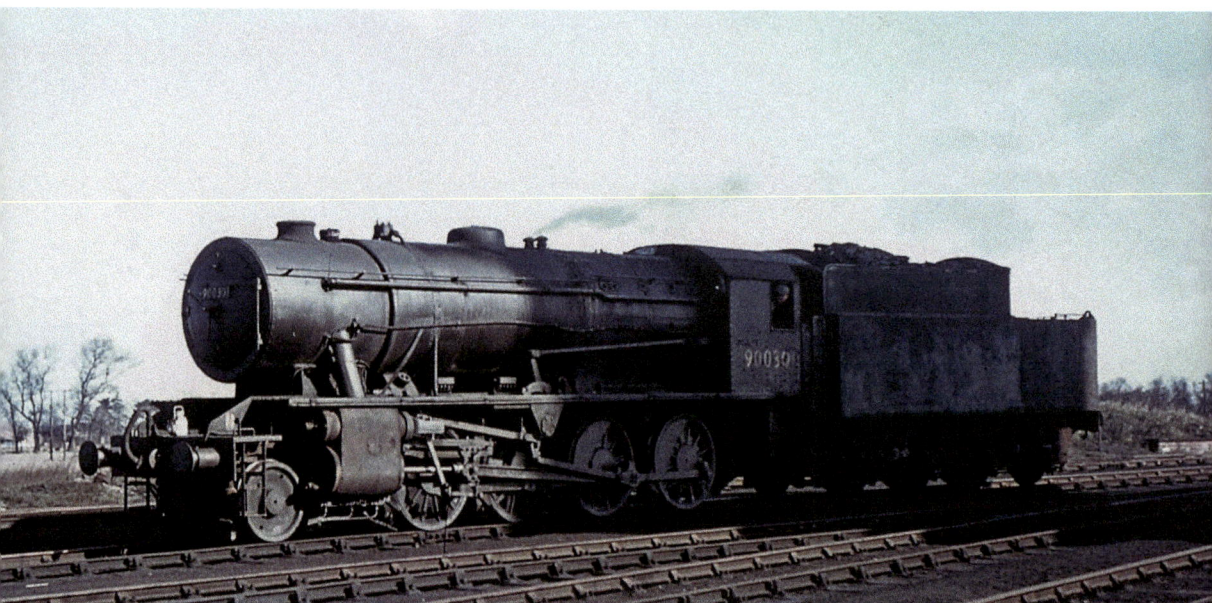

18 March 1967. WD 2-8-0 No. 90030 drifts onto shed at Goole.

WD 2-8-0 No. 90030 runs onto shed at Goole in early 1967. No. 90030 had been allocated to the shed since November 1965 but would be withdrawn in April. It was to join the lines of withdrawn locomotives awaiting removal for scrap to Draper's of Hull. The depot itself was only a few months away from closing to steam at the end of May 1967.

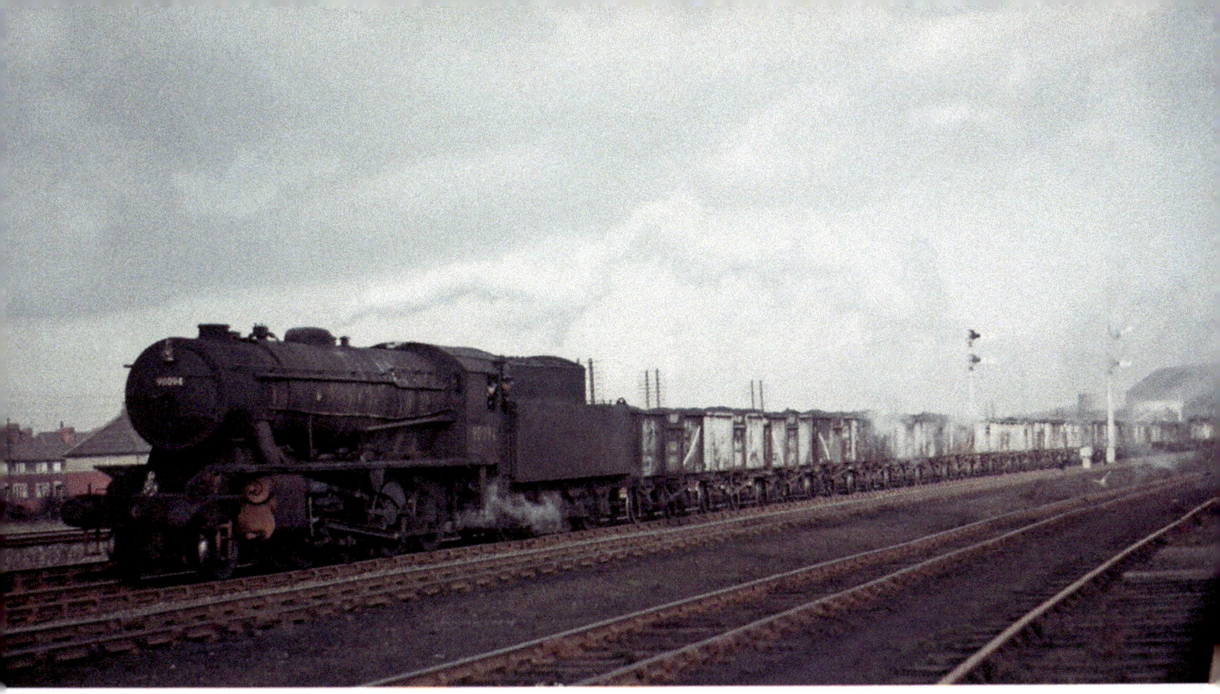

April 1967. WD 2-8-0 No. 90094 heads for Goole.

WD 2-8-0 No. 90094 on a string of loaded coal wagons heads for Goole. The WD had been based at Goole MPD since late in 1958 and must have been one of the last working locomotives at the depot. It appears to have been withdrawn officially after the depot had been closed to steam and would make its way to Draper's of Hull for scrapping like so many other Goole engines.

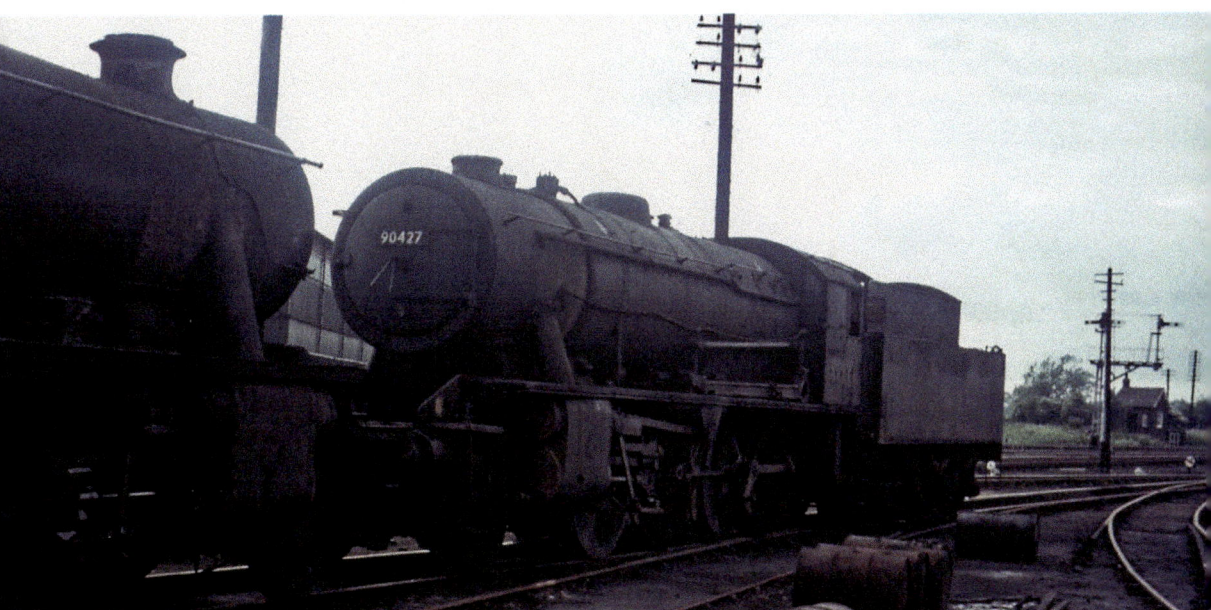

June 1965. WD 2-8-0 No. 90427 from Goole.

WD 2-8-0 No. 90427 had been a Hull-allocated engine from 1950 and then transferred to Goole MPD in 1963. It was to remain there until the depot closed to steam at the end of May 1967. It was seen coupled to an LMS 8F 2-8-0 at Hull Botanic Garden MPD presumably en route to Darlington for overhaul. LMS 8Fs were, at the time, the most common class to be repaired at Darlington.

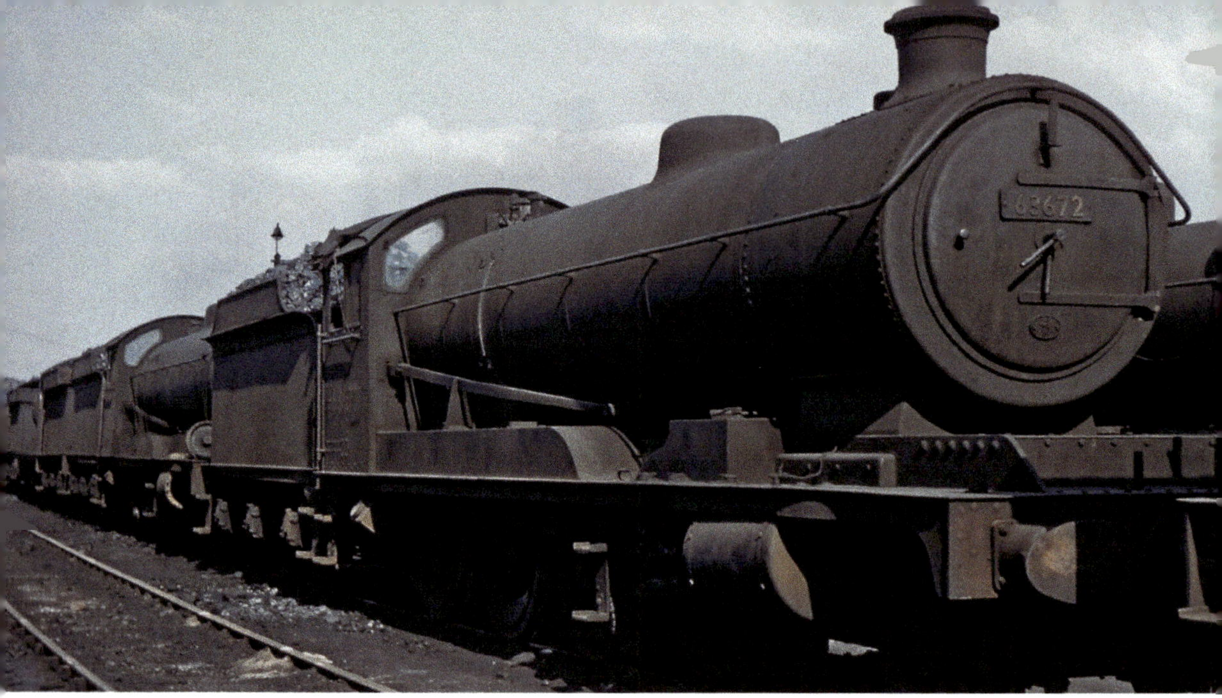

June 1960. LNER O4/8 2-8-0 No. 63672 on Mexborough MPD.

LNER O4/8 2-8-0 No. 63672 had been built as an O4/3 in October 1917 by Robert Stephenson and Hawthorn for the War Department. It was eventually bought by the LNER in June 1924 and worked from several ex-GCR depots before arriving at Mexborough MPD in the summer of 1946. It was to remain a Mexborough locomotive until December 1961 and during its time there in April 1957 it was rebuilt into its final form as an O4/8 with a B1 type boiler and cab. After its time at Mexborough it moved to Retford Thrumpton MPD and was finally withdrawn in December 1963.

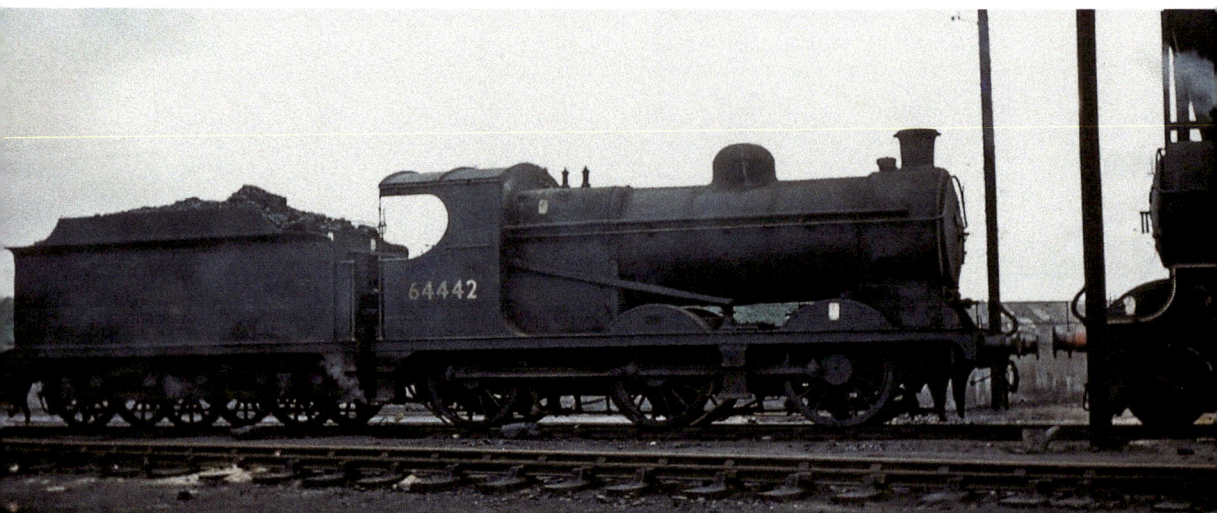

September 1960. LNER J11/3 0-6-0 No. 64442 on Mexborough MPD.

LNER J11/3 0-6-0 No. 64442 had been built by the Great Central Railway at its Gorton Works in October 1908. As part of Thompson's standardisation programme for the LNER, J11s were chosen for modification and No. 64442 was rebuilt in May 1947. Long travel piston valves were fitted, which meant the superheated boiler had to be pitched higher. In turn this required modification to the cab, front plate and the chimney. No. 64442 had been allocated to Mexborough MPD on several different occasions in its fifty-three-year career and would be withdrawn from there in September 1962.

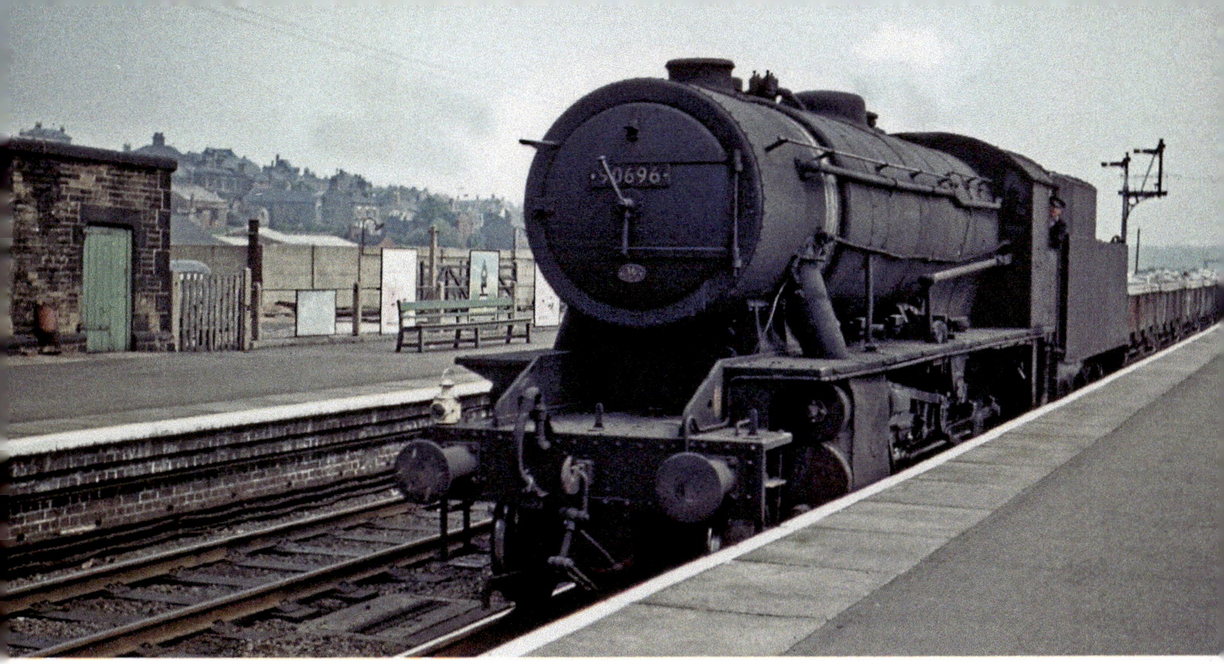

May 1963. WD 2-8-0 No. 90696 at Mexborough station.

WD 2-8-0 No. 90696 was built in December 1944 by Vulcan Foundry as WD No. 79242. After being shipped to Belgium it was put to work on the SNCF, the French State railway, but was back in the UK and on loan to the LNER at Mexborough MPD by March 1947. In 1951 it became British Railways No. 90696 and spent much of its career at Doncaster MPD with occasional stays at March depot. In February 1960 it moved to Frodingham MPD and remained at that depot until its withdrawal in July 1963.

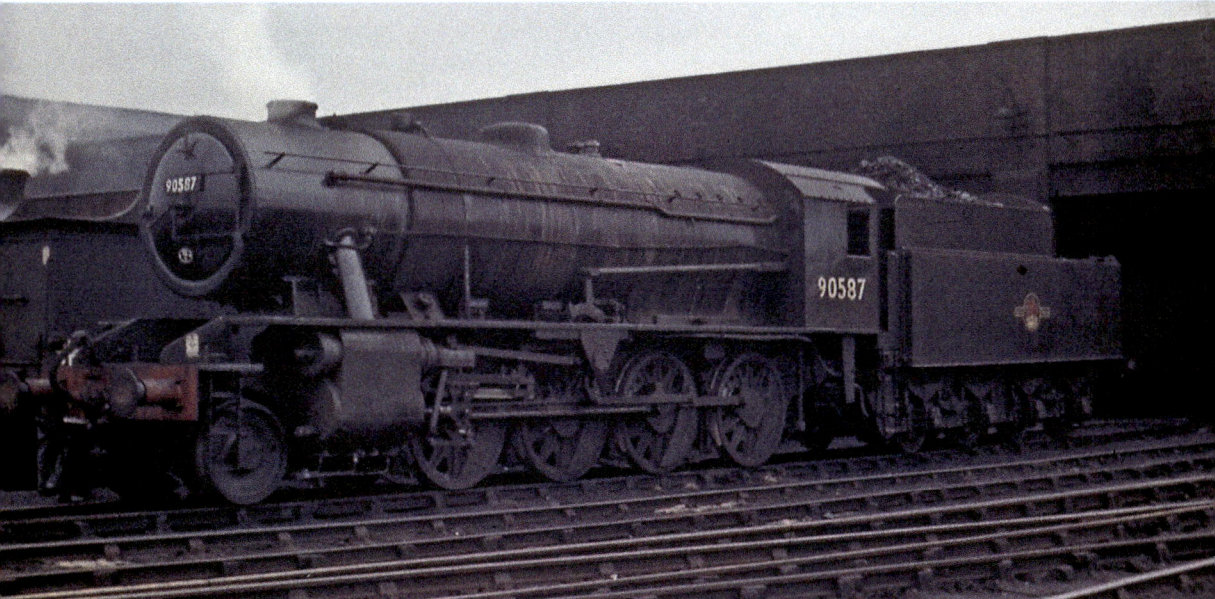

May 1963. WD 2-8-0 No. 90587 on Mexborough MPD.

WD 2-8-0 No. 90587 was still looking reasonably tidy after its general overhaul in Doncaster Plant earlier in the year. It was a long-term resident of Mexborough MPD having been allocated there in 1948 and was to stay loyal to the depot until its closure in February 1964. The locomotive then moved to Staveley and was finally withdrawn from Langwith Junction in November 1965.

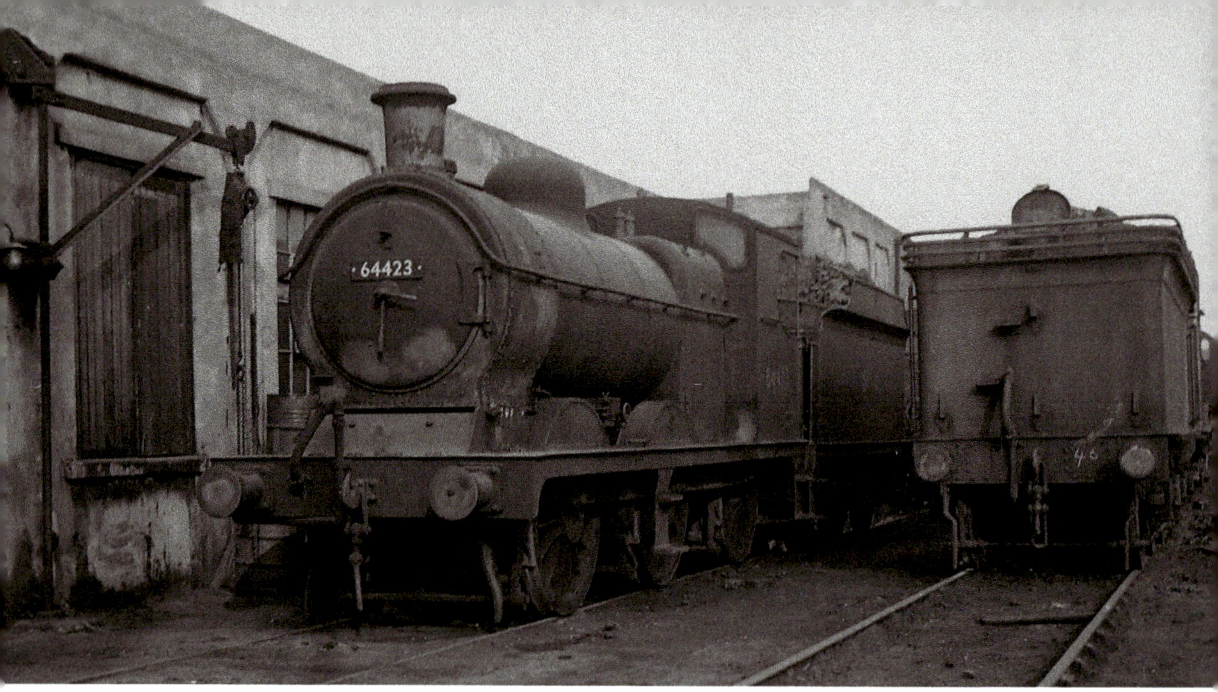

20 March 1960. LNER J11/5 0-6-0 No. 64423 on Frodingham MPD.

LNER J11/5 0-6-0 No. 64423 had entered service for the Great Central Railway in October 1907 and had few changes made to the original design. In 1938 it had acquired a superheated boiler and a lower height chimney to become part of the J11/5 sub class. It had become a Frodingham-allocated locomotive in late 1959 and was withdrawn from the depot in March 1962. By the end of the year the whole class would be rendered extinct.

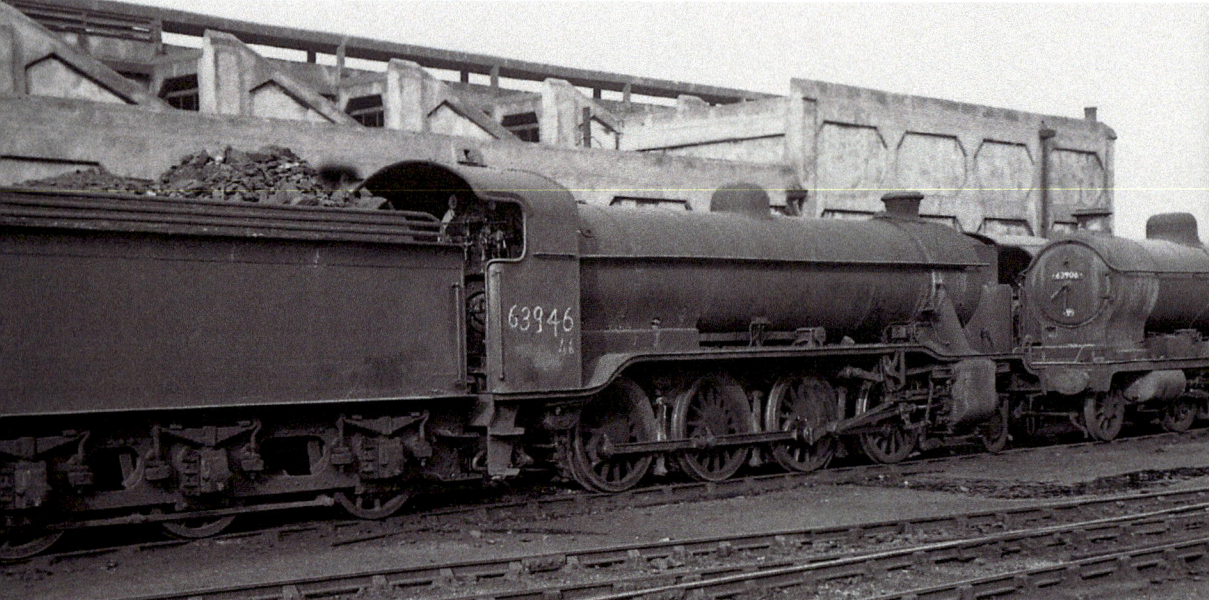

20 March 1960. LNER O2 2-8-0 No. 63946 on Frodingham MPD.

LNER O2 2-8-0 No. 63946 had been built at Doncaster Plant a year after the formation of the LNER but retained all the Great Northern Railway features of the earlier members of the class, most notably the rather spartan-looking cab. Since April 1953, this locomotive had been allocated to Grantham MPD and was a regular visitor to Frodingham, working the heavy iron ore trains from the High Dyke branch just south of Grantham.

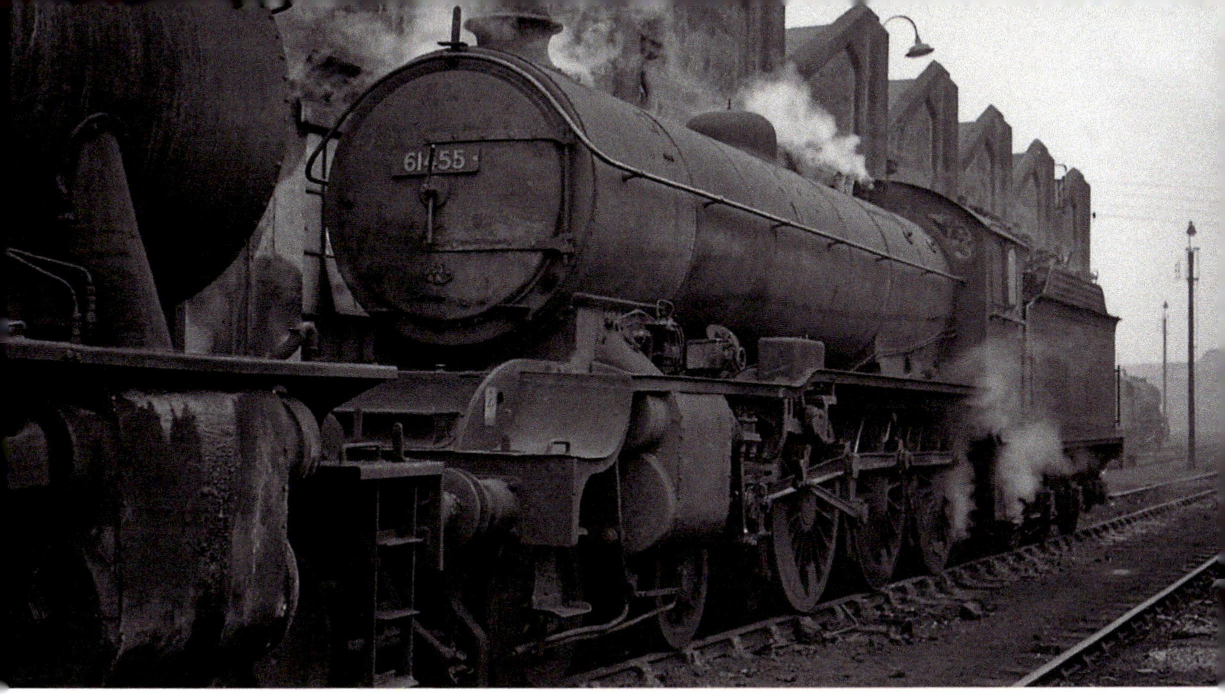

September 1962. LNER B16/2 4-6-0 No. 61455 at Frodingham MPD.

LNER B16/2 4-6-0 No. 61455 had been built for the LNER to a North Eastern design in October 1923, some ten months after the grouping of the railway companies into the Big Four. No. 61455 was one of six of the class to be rebuilt by Gresley with Walschaerts valve gear on the outside cylinders and conjugated valve gear on the middle cylinder. It was to remain a York-based locomotive until withdrawal in September 1963.

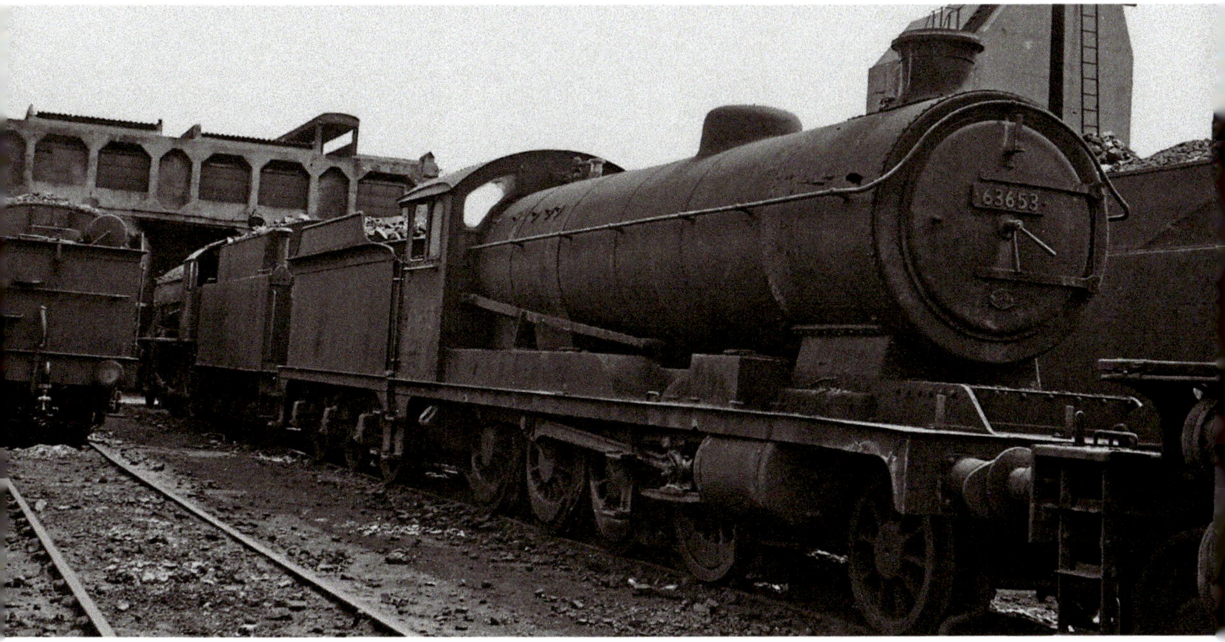

July 1963. LNER O4/8 2-8-0 No. 63653 on Frodingham MPD.

LNER O4/8 2-8-0 No. 63653 had been built for the War Department in April 1918 and bought by the LNER in 1924. In February 1946 it was rebuilt with B1 type boiler and cab as an O4/8 and was reallocated to Frodingham in March 1950. It was to transfer to Doncaster MPD for the last six months of its career.

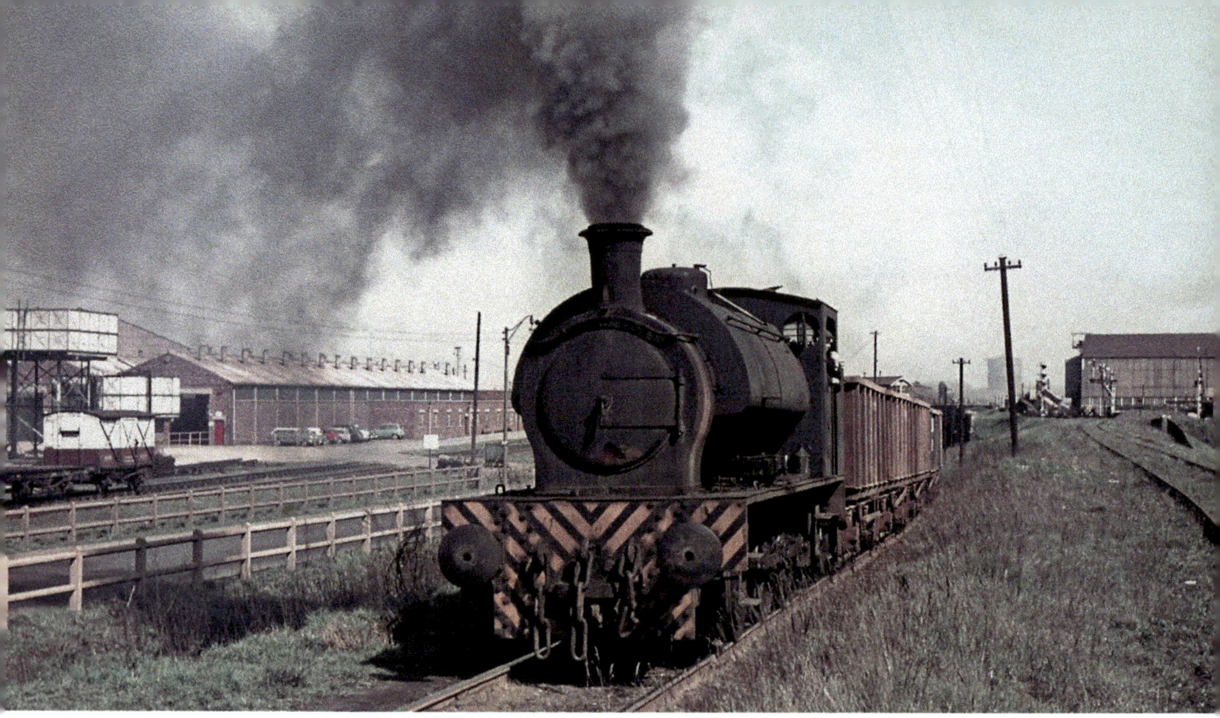

17 April 1968. 0-6-0ST RT&B No. 17 on the North Lindsey Light Railway Scunthorpe.

0-6-0ST Hudswell Engine Company (class 48150) No. 17 was another locomotive built for Richard Thomas and Baldwins Redbourn Ironworks, Scunthorpe. It is seen heading back to the Crosby Mines along the North Lindsey Light Railway. (L. Flint)

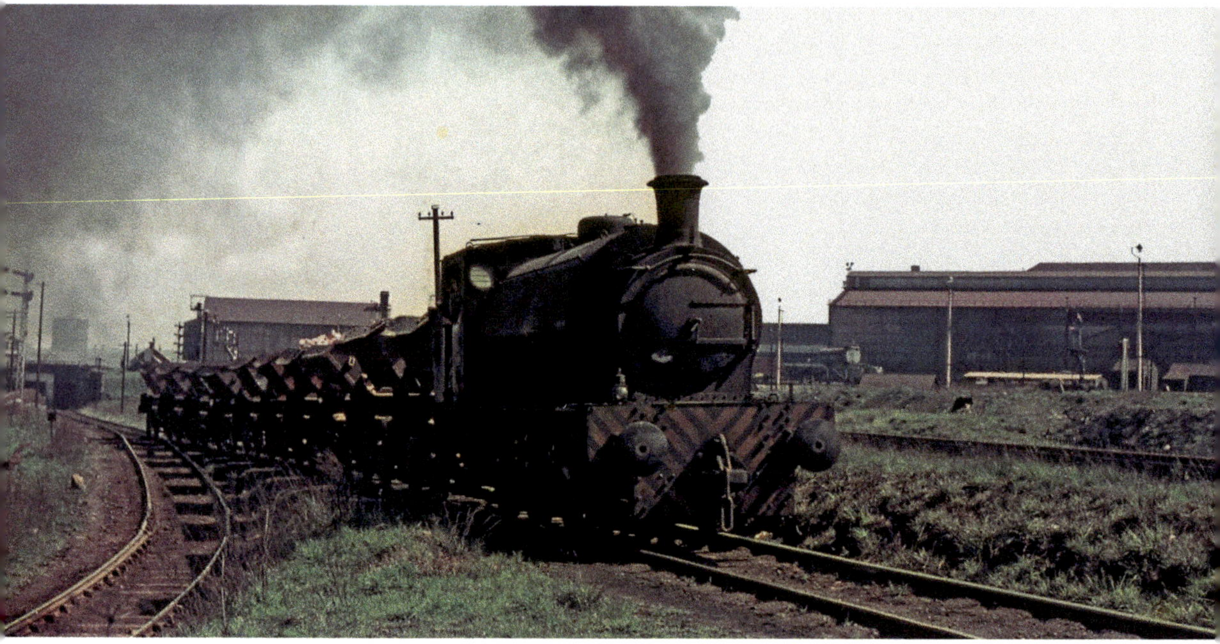

17 April 1968. 0-6-0ST RT&B No. 15 on the North Lindsey Light Railway Scunthorpe.

0-6-0ST Hudswell Engine Company (class 48150) No. 15 was another locomotive built for Richard Thomas and Baldwins Redbourn Ironworks, Scunthorpe. It is seen heading back to the Crosby Mines along the North Lindsey Light Railway. (L. Flint)

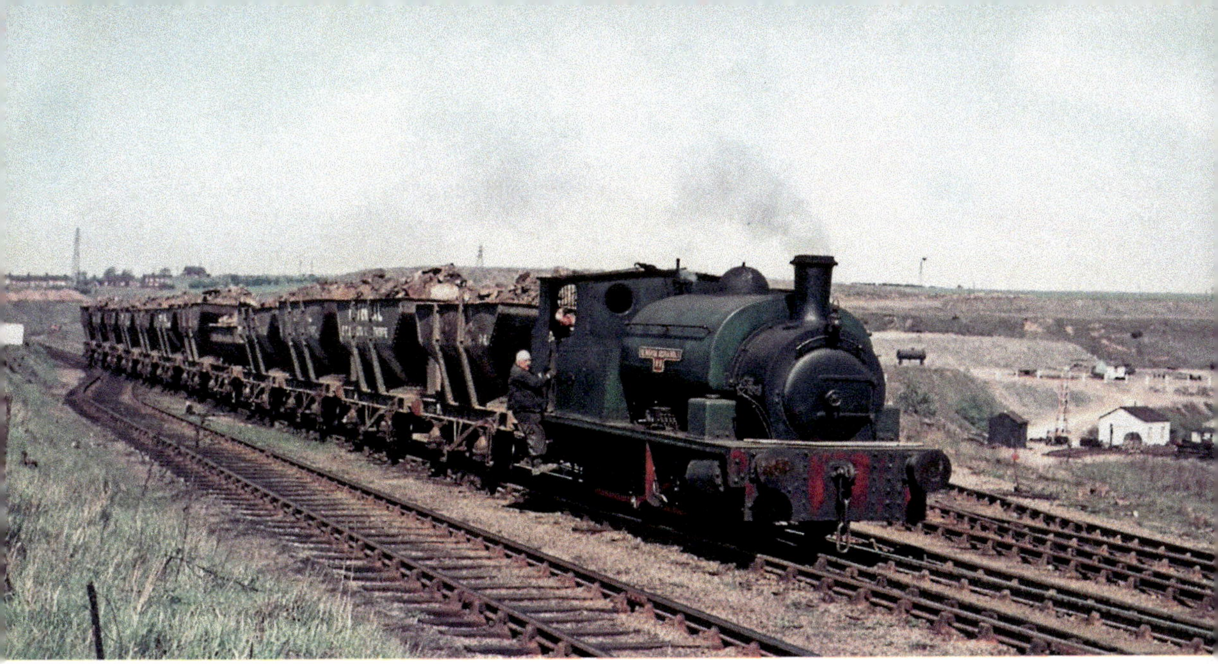

17 April 1968. 0-6-0ST *Renishaw No. 6* at Crosby Mines in Scunthorpe.

0-6-0ST *Renishaw No. 6* was built in November 1919 by Hudswell Clark (works number 1366). As No. 16, it was new to the Appleby Iron Co. Ltd, Appleby Ironworks, Scunthorpe, Lincolnshire. The company became part of the Appleby-Frodingham Steel Co. Ltd in October 1934. It was sold in 1952 to Renishaw Ironworks in Derbyshire and was named Renishaw Ironworks No. 6. In 1961, it was sold again and returned to Scunthorpe to work at the Midland Ironstone Mines at Crosby, hauling iron ore as seen here to the steel furnaces of Scunthorpe. (L. Flint)

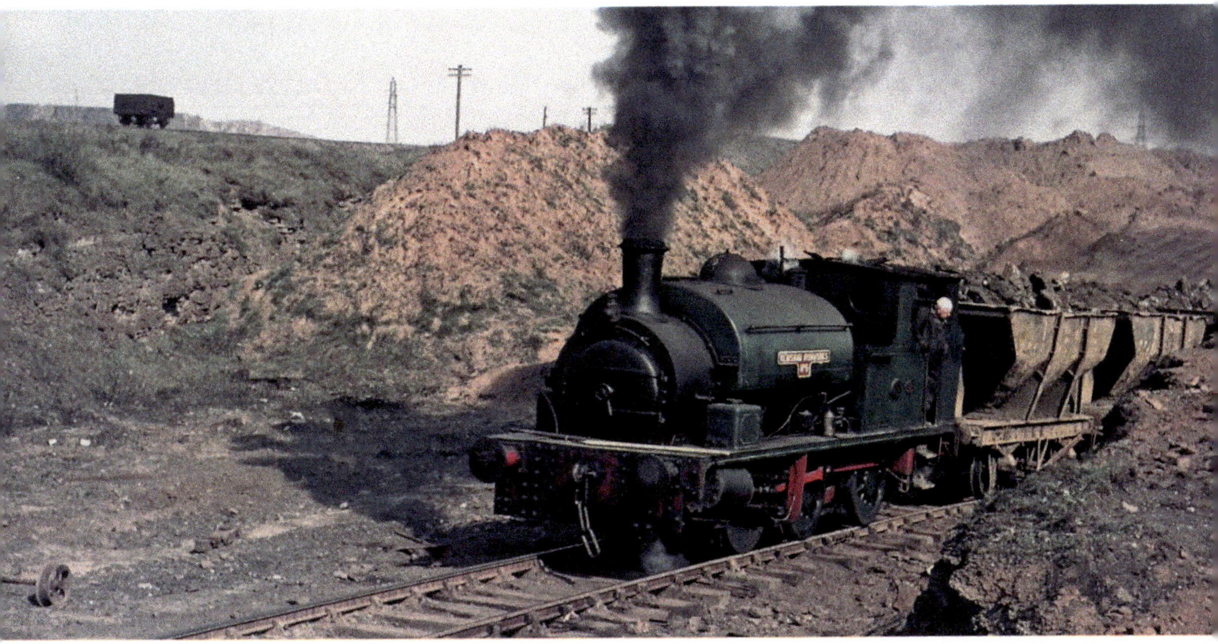

17 April 1968. 0-6-0ST *Renishaw No. 6* climbs out of Crosby Mines in Scunthorpe.

0-6-0ST *Renishaw No. 6* hard at work in the Midland Ironstone Mines at Crosby hauling iron ore to the steel furnaces of Scunthorpe. When operations at the mine ceased in 1969, the locomotive was purchased for preservation and is now at the Tanfield Railway in County Durham. (L. Flint)

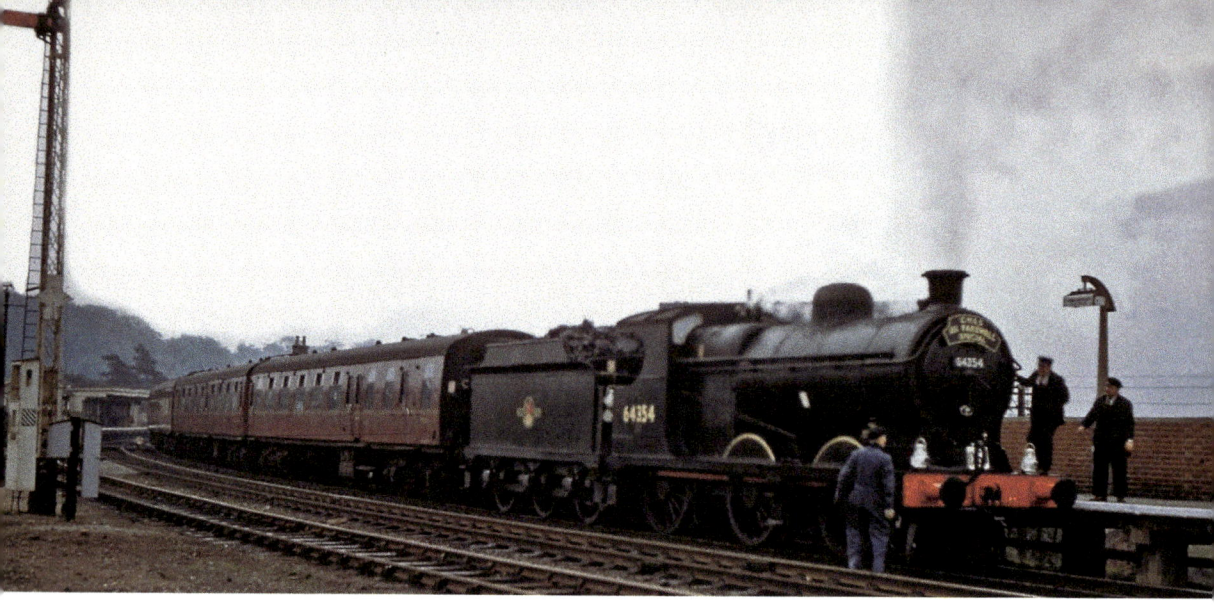

3 October 1962. LNER J11/3 0-6-0 No. 64354 at Gainsborough Lea Road.

LNER J11/3 0-6-0 No. 64354 waits at Gainsborough Lea Road station with the J11 Farewell Special, which had been organised by the Gainsborough Model Railway Society to mark the demise of this successful, long-lived Great Central Railway design. The locomotive had been built in 1903 at Gorton Works and rebuilt in 1943 as a J11/3, which included the fitting of long travel piston valves and associated modifications to the boiler mountings. It had spent its career working around the Great Central area with its last two years allocated to Retford MPD. At the time of this special working it was the sole surviving working member of the class and was withdrawn immediately afterwards. (L. Flint)

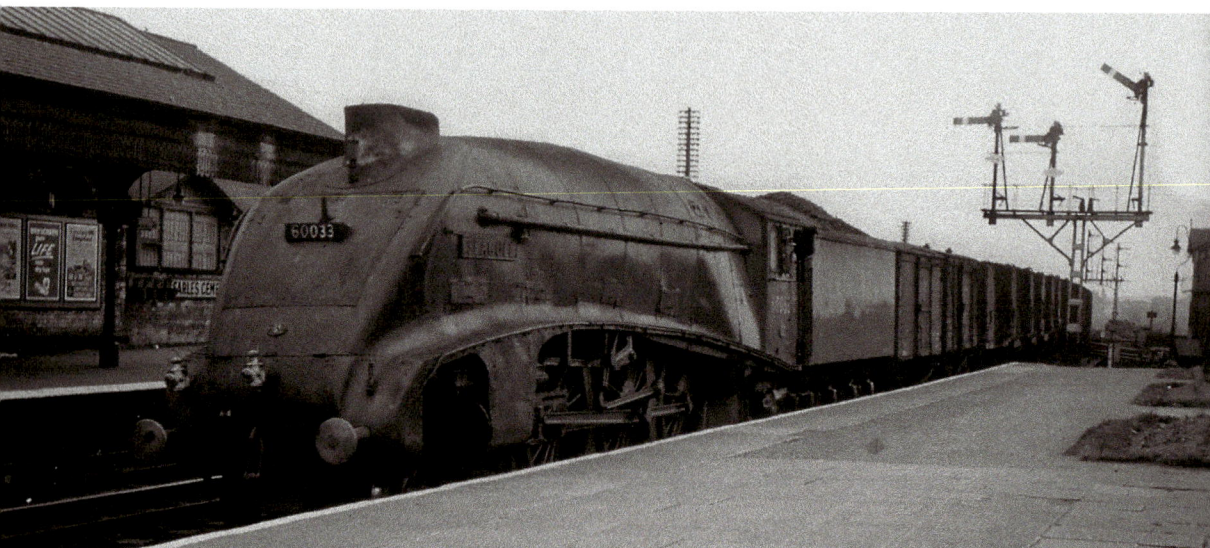

May 1961. LNER A4 4-6-2 No. 60033 storming through Retford.

LNER A4 4-6-2 No. 60033 *Seagull*, with a well-filled tender, whistles as it speeds northbound through Retford at the head of an express Class C fitted van train. The train was the daily Scotch Goods, which left London just before three o'clock in the afternoon and ran at express speeds to Niddrie, outside Edinburgh. The locomotive was a long-term resident of Kings Cross shed, having been allocated to only one other depot in its career, but was to be an early casualty to the rapid dieselisation of East Coast Main Line services and was withdrawn in December 1962. (L. Flint)

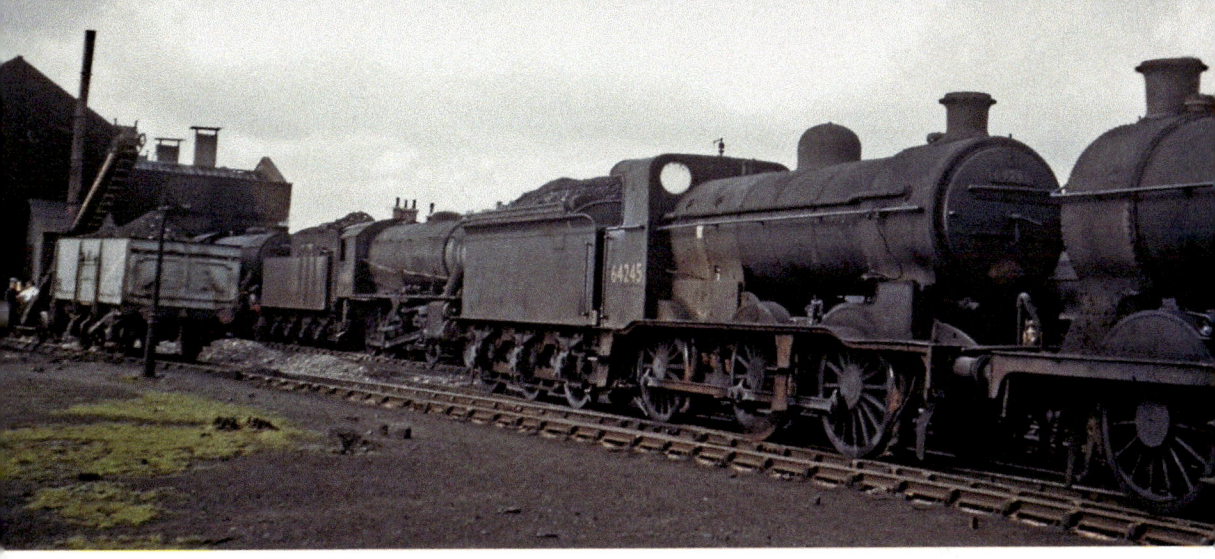

June 1961. LNER J6 0-6-0 No. 64174 on Retford Thrumpton MPD.

LNER J6 0-6-0 No. 64174 waits for its next duty at its home shed of Retford Thrumpton (G.C.). It had been a Retford engine for the last six years but it was to be withdrawn a couple of months after this photograph. It had been built at Doncaster in 1911 with the design attributed to both H. Ivatt and N. Gresley, marking an important milestone for the Great Northern Railway as one chief mechanical engineer handed over the reins to his successor.

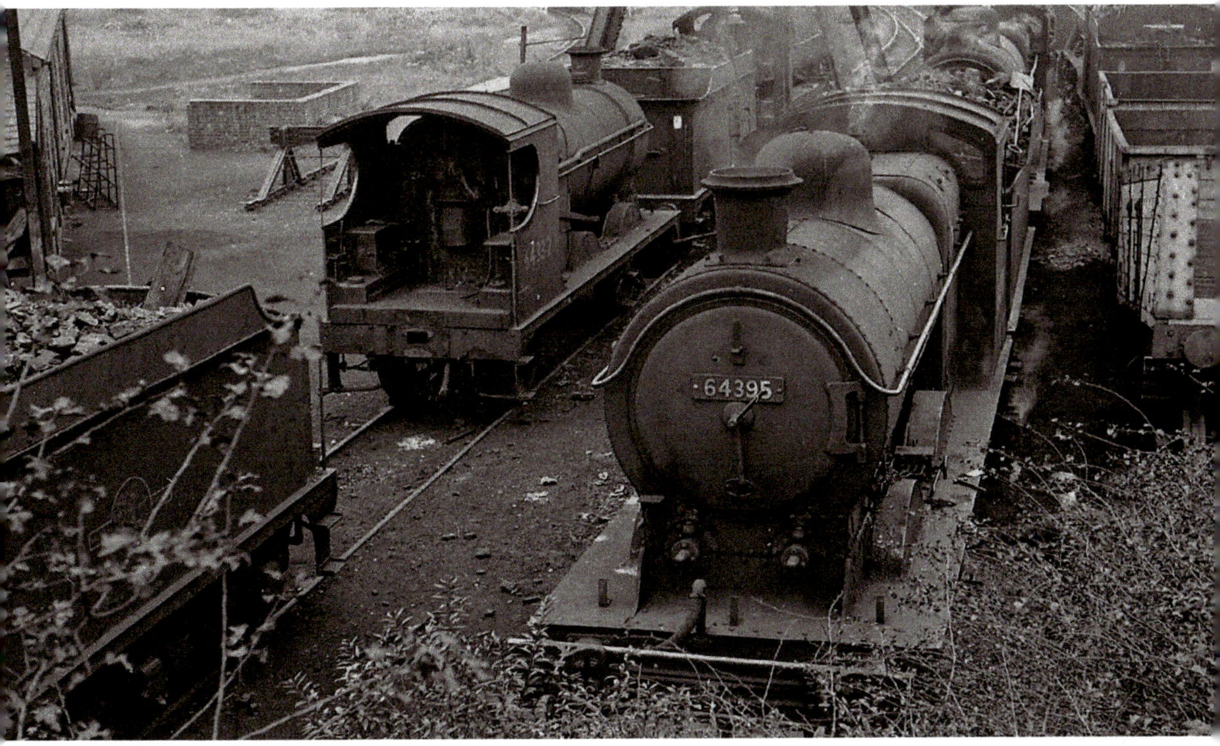

July 1961. LNER J11/3 No. 64395 on Retford Thrumpton (G.C.) MPD.

LNER J11/3 Nos 64395 and 64332 were more familiar residents of the old Great Central Railway shed than the J6s but they lasted only a few months longer in service. No. 64395 would be withdrawn at the end of January 1962 and No. 64432 would follow in September of 1962. Visible in this view of the depot is the shear legs lifting gantry, which were a feature of Great Central depots.

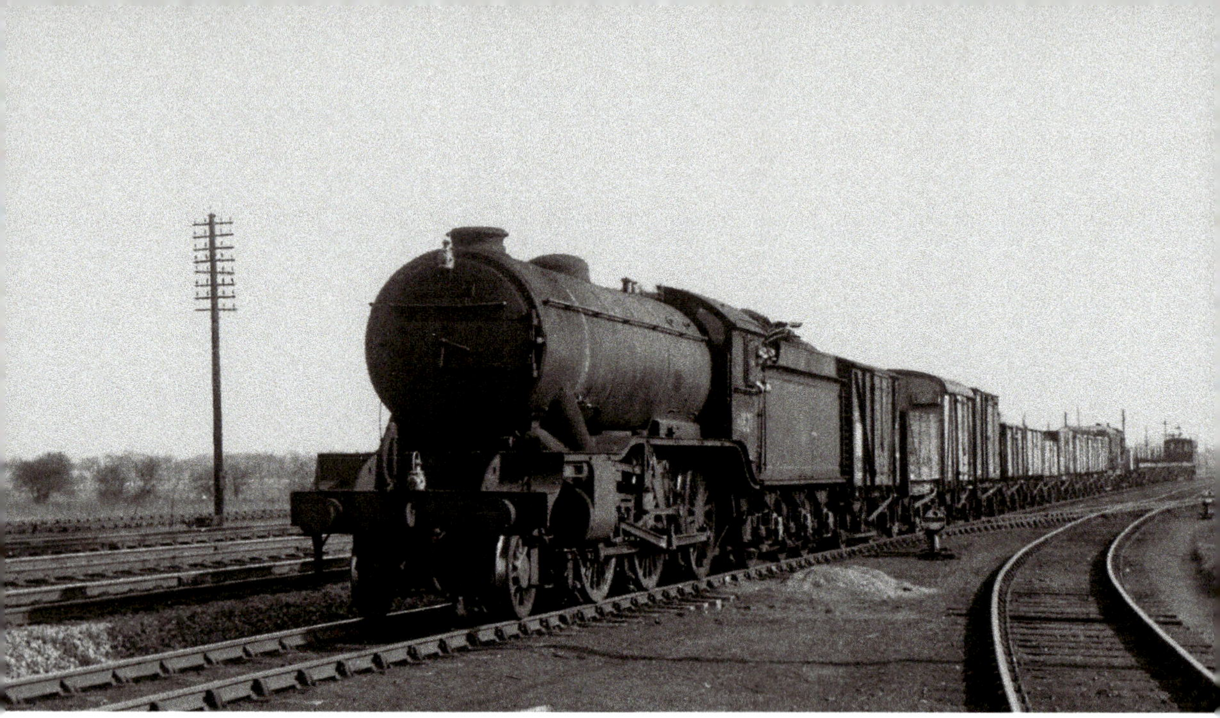

July 1961. LNER K3 2-6-0 No. 61821 near Retford.

LNER K3 2-6-0 No. 61821 heads south through Retford with a mixed goods train. The locomotive had been built in 1924 and had been allocated to Colwick for a number of years. It was to be withdrawn in September 1962 and by the end of that year the whole class was to be taken out of service.

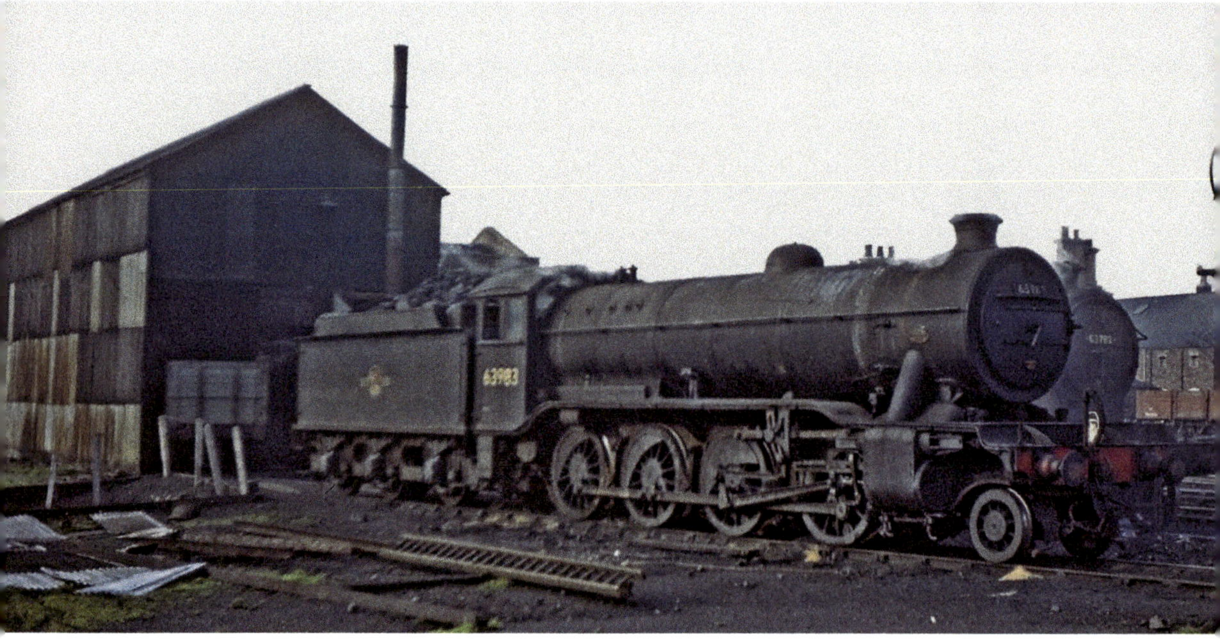

October 1961. LNER O2 2-8-0 No. 63983 on Retford MPD.

LNER O2 2-8-0 No. 63983 is prepared for its next duty in the dilapidated surroundings of Retford G.C. MPD. The rusting corrugated-iron building and scattered scrap and detritus strewn about creates a sadly neglected air and an insight into the everyday working environment of those involved in the railway industry at that time.

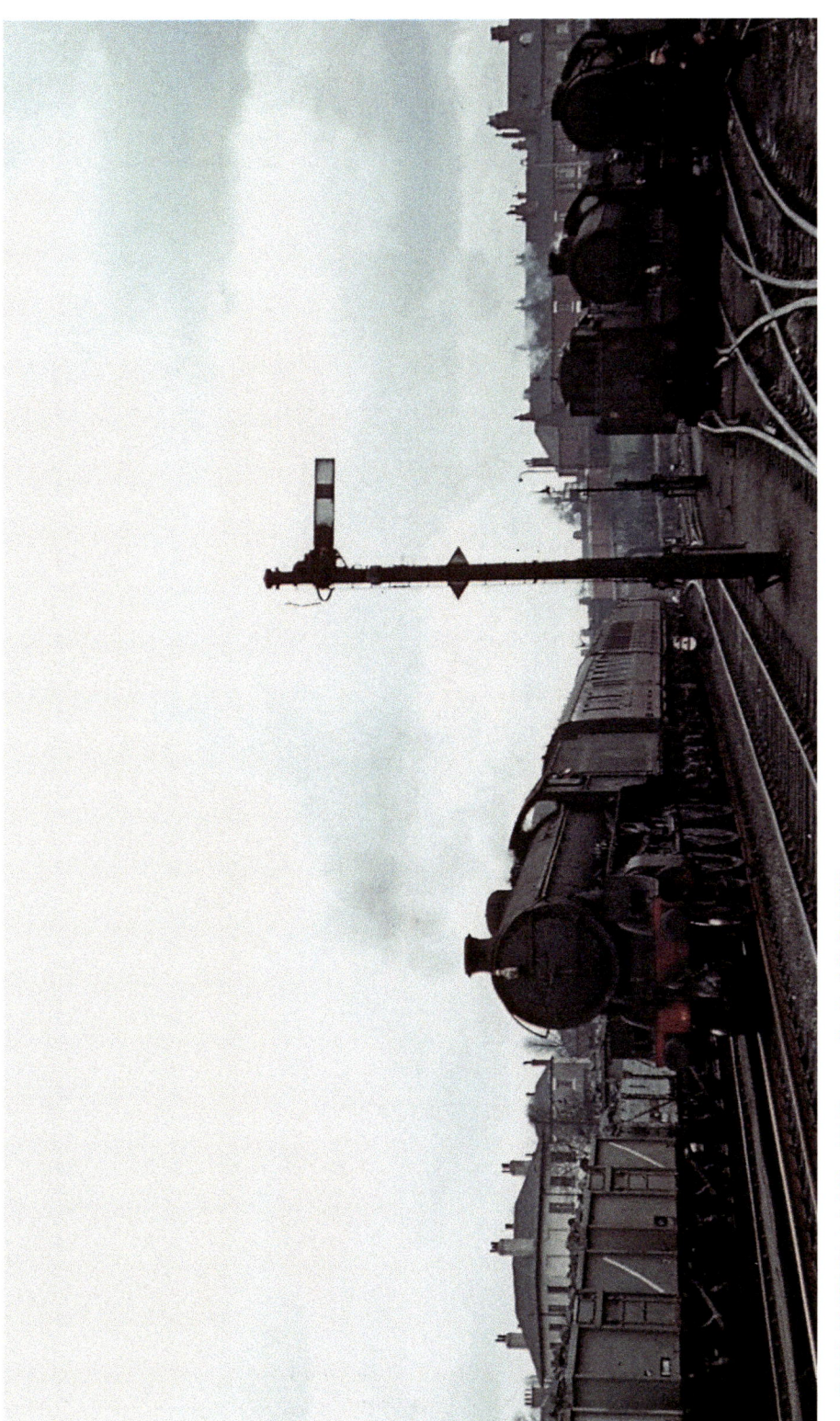

April 1962. LNER B1 4-6-0 No. 61145 passes Retford G.C. MPD.

LNER B1 4-6-0 No. 61145 heads a local passenger train passed the great Central depot in Retford. No. 61145 was still a Doncaster-allocated locomotive at this time but was soon to move on to Immingham before finally ending its days in January 1966 at Colwick MPD.

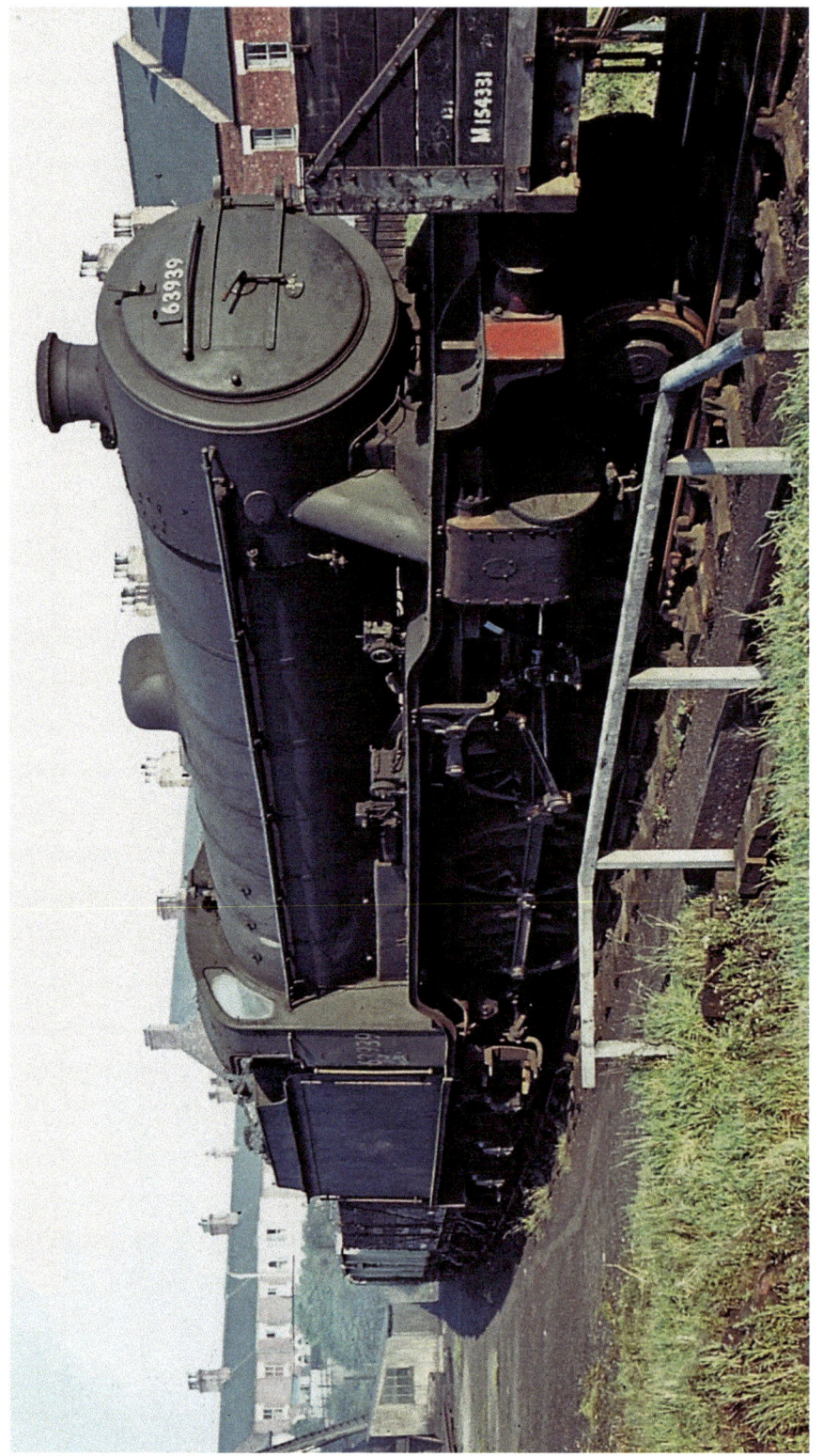

July 1962. LNER O2 2-8-0 No. 63939 on Retford G.C. MPD.

LNER O2 2-8-0 No. 63939 had been built at Doncaster in December 1923 and had been allocated to Retford G.C. MPD since January 1960. It was seen stored at the depot in the summer of 1962, fully coaled and looking quite presentable. It was to return to active service again for a short while, being finally withdrawn in September 1963 and cut up in February 1964.

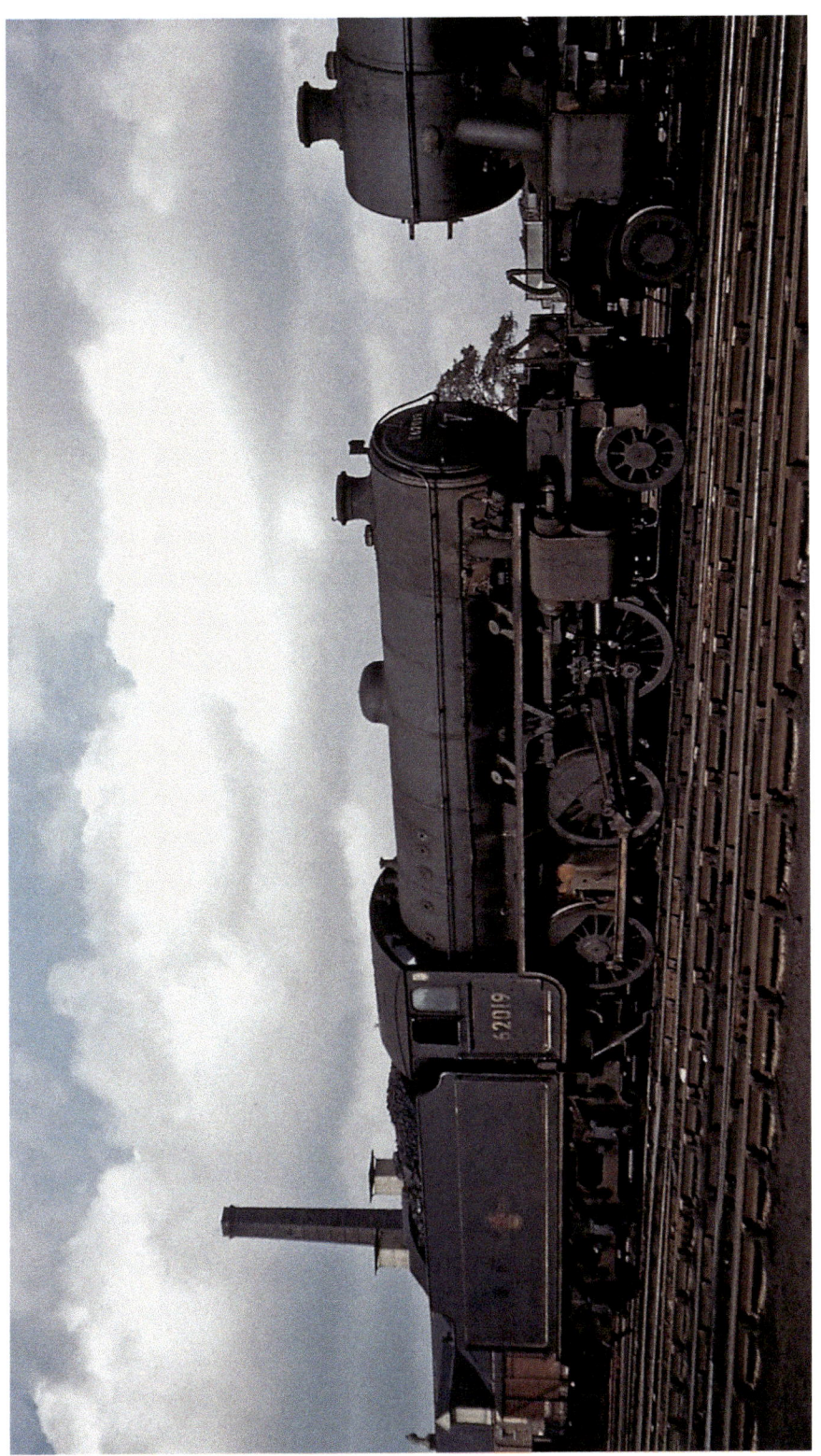

July 1962. LNER K1 2-6-0 No. 62019 at Retford.

LNER K1 2-6-0 No. 62019 was new to British Railways having been built in July 1949 by the North British Locomotive Works in Glasgow. It had spent much of its short working life on Great Eastern territory and escaped the wave of dieselisation there by finding sanctuary at Retford in September 1961. It was to survive a visit to Doncaster Plant in November 1962, where it had a heavy casual overhaul and remained working at Retford until withdrawal in July 1964.

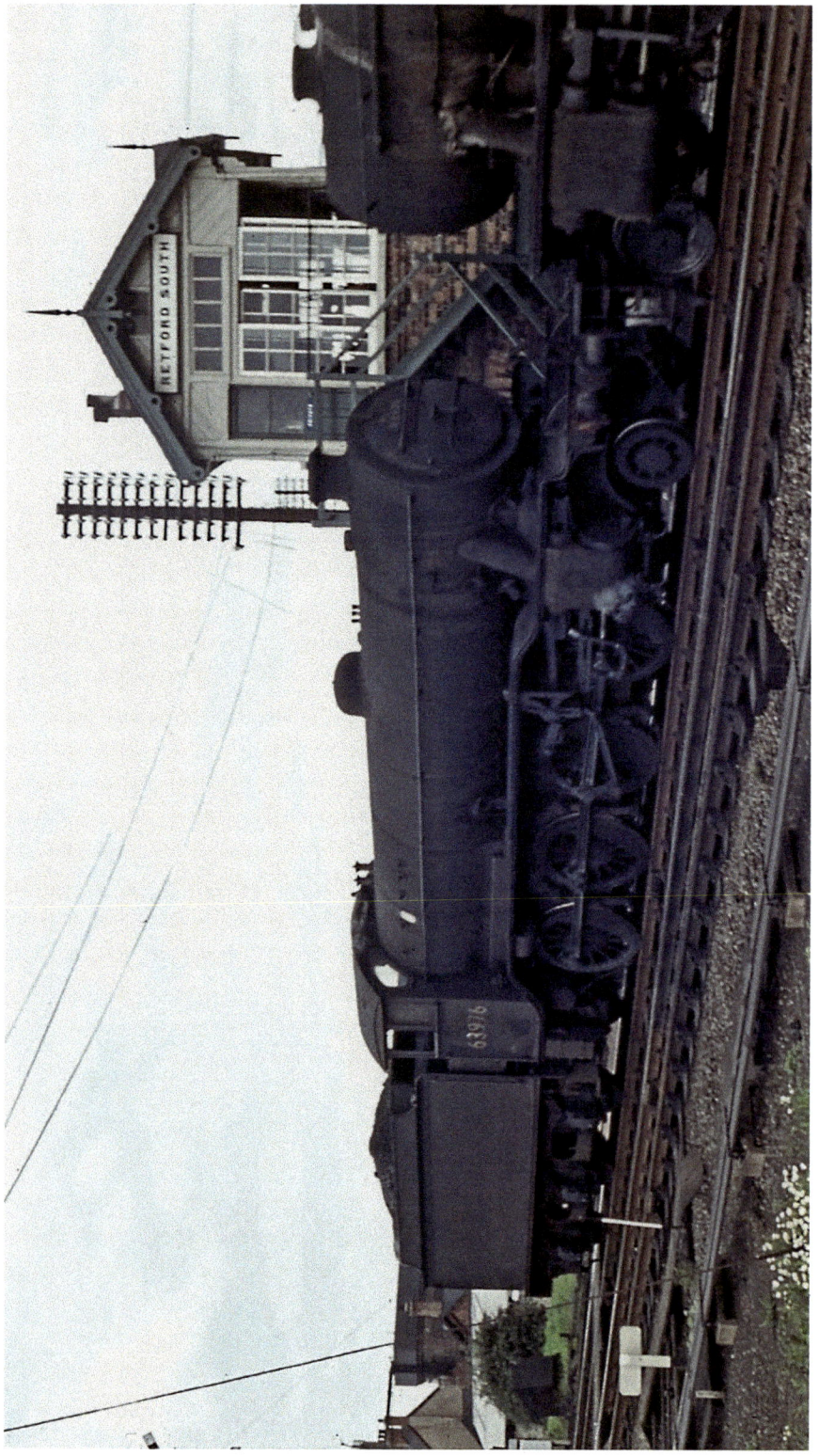

June 1963. LNER O2 2-8-0 No. 63976 heads past Retford South.

LNER O2 2-8-0 No. 63976, with a WD 2-8-0, runs light engine back to Retford G.C. MPD along the old Great Central route. They are at the flat crossing with the East Coast Main Line, which was controlled by Retford South signal box. In 1965, the crossing was replaced by a dive which took the G.C. route below the ECML and changed this view completely.

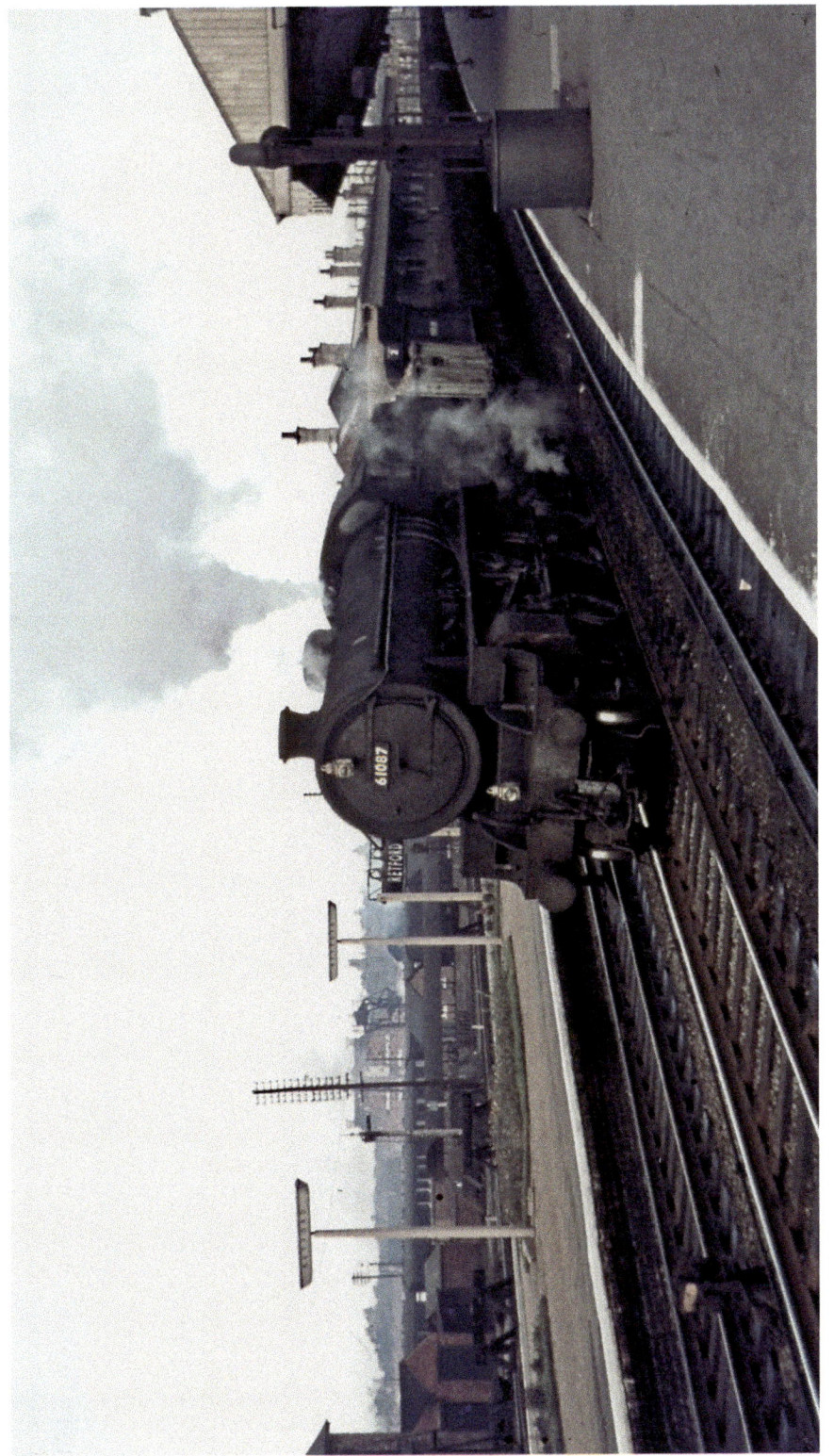

June 1964. LNER B1 4-6-0 No. 61087 trundles through Retford station.

LNER B1 4-6-0 No. 61087 trundles through Retford station with a long permanent way train. No. 61087 had been built for the LNER by the North British Locomotive Company in October 1946 and had been a Doncaster-based engine since February 1949. As shown in an earlier illustration, No. 61087 had received what was to be its last general overhaul at the Plant in January 1963 and it would remain a familiar sight around Doncaster until withdrawal in December 1965.

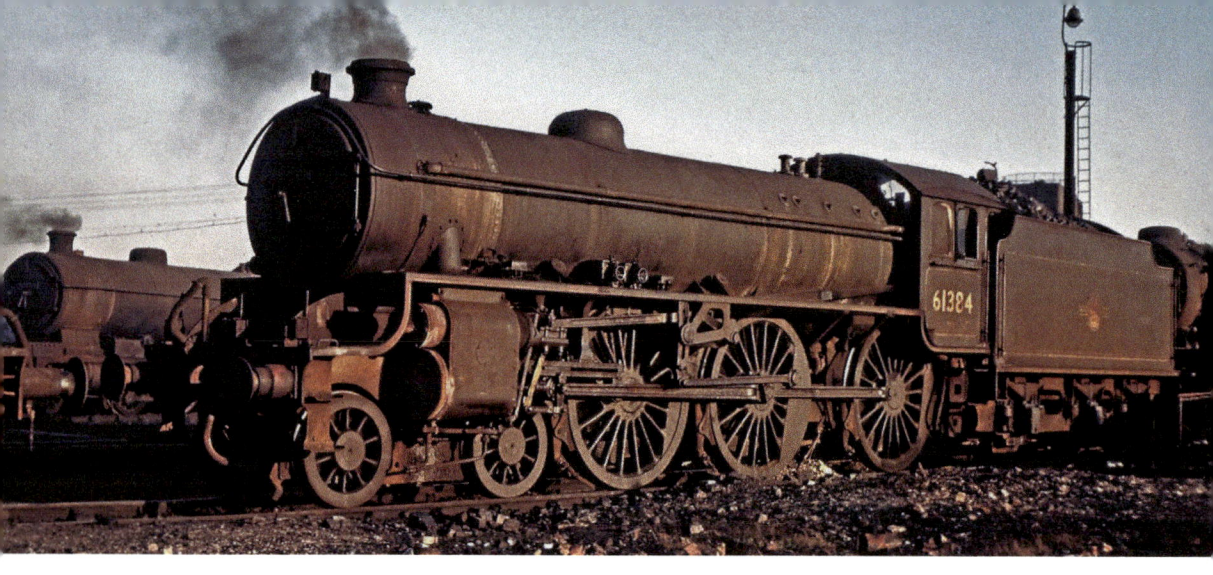

28 March 1964. LNER B1 4-6-0 No. 61384 at Retford MPD.

LNER B1 4-6-0 No. 61384 basks in the afternoon sunshine at Retford MPD. It had been built by the North British Locomotive Company of Glasgow for British Railways in October 1951 and was a resident of Immingham MPD at the time of the photograph. In April 1965 it was reallocated to Retford for a few months before returning to Immingham and its withdrawal in January 1966.

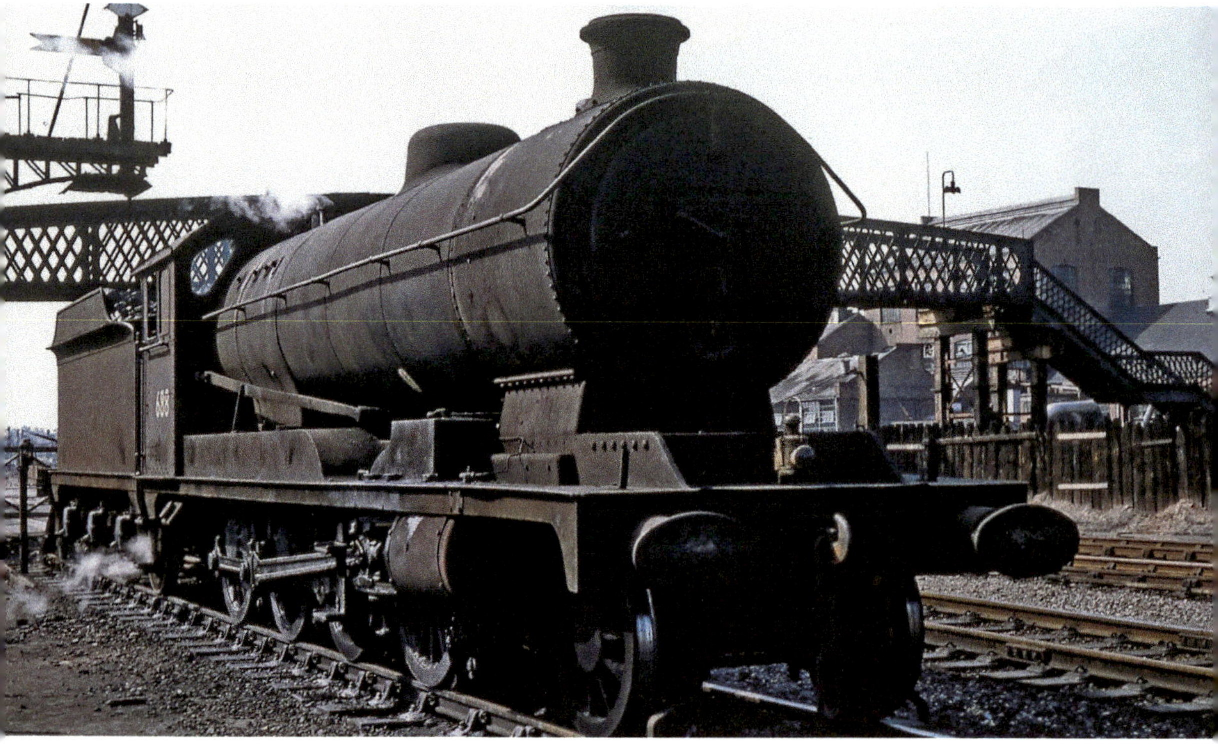

July 1964. LNER O4/8 2-8-0 No. 63688 at Retford.

LNER O4/8 2-8-0 No. 63688 had been built for the Railway Operating Division of the Royal Engineers by Robert Stephenson and Hawthorn of Newcastle in February 1918. It was bought by the LNER in 1924 and since that time was allocated to either Doncaster or Retford depots. In November 1956 it was rebuilt as an O4/8 with a B1 type boiler and cab. It had one more move back to Doncaster in June 1965 before withdrawal from there three months later.